Feline Behaviour and Welfare

This book is dedicated to my wife and my children, grandchildren and great-grandchildren: too many to name but never far from my thoughts.

Feline Behaviour and Welfare

Andrew F. Fraser

www.cabi.org

CABI is a trading name of CAB International

CABI
Nosworthy Way
Wallingford
Oxfordshire OX10 8DE
UK

Tel: + 44 (0)1491 832111
Fax: + 44 (0)1491 833508
E-mail: info@cabi.org
Website: www.cabi.org

CABI
875 Massachusetts Avenue
7th Floor
Cambridge, MA 02139
USA

Tel: + 1 800 552 3038 (toll free)
Tel: +1 (0)617 395 4051
E-mail: cabi-nao@cabi.org

A catalogue record for this book is available from the British Library, London, UK.

Library of Congress Cataloging-in-Publication Data

Fraser, Andrew Ferguson.
Feline behaviour and welfare / by Andrew F. Fraser.
p. ; cm.
Includes bibliographical references and index.
ISBN 978-1-84593-927-4 (pbk. : alk. paper) -- ISBN 978-1-84593-926-7 (hbk : alk. paper)
1. Felidae--Behavior. 2. Cats--Behavior. I. C.A.B. International. II. Title.
[DNLM: 1. Cats. 2. Animal Welfare. 3. Behavior, Animal. SF 446.5]

QL737.C23F744 2012
599.74--dc23

2012004662

ISBN: 978 1 84593 927 4 (pbk)
978 1 84593 926 7 (hbk)

Commissioning editor: Sarah Hulbert
Editorial assistant: Alexandra Lainsbury
Production editor: Tracy Head

Typeset by SPi, Pondicherry, India
Printed and bound by Gutenberg Press Limited, Tarxien, Malta

Contents

About the Author

Andrew Ferguson Fraser obtained his veterinary education in Glasgow, Scotland and Guelph, Canada. He was a veterinary practitioner for 12 years before entering the academic field, teaching at the Royal Dick School of Veterinary Medicine, Edinburgh University for 12 years. He moved to Canada where he remained until retiring as Professor of Surgery (Veterinary) in the Medical School of Memorial University of Newfoundland. In retirement, he has continued his interest in veterinary ethology, in which field he was a pioneer and the founding editor-in-chief of the journal *Applied Animal Behaviour Science*, continuing in that role for 16 years.

The author's ethological books include *Reproductive Behaviour in Ungulates* (1968, Academic Press), *Farm Animal Behaviour* (1974 and 1980, Bailliere and Tindall), *Farm Animal Behaviour and Welfare* with D. Broom (1990, Bailliere and Tindall), *The Behaviour of the Horse* (1992, CAB International), *Domestic Animal Behaviour and Welfare* with D. Broom (2007, CAB International) and *The Behaviour and Welfare of the Horse* (2010, CAB International).

He has been a cat owner throughout his life and has long been involved with cats in his veterinary work. He became acutely aware of the need for improved feline welfare after service with animal shelter organizations and following visits to several areas of the world where wild felines are in danger of extinction from excessive hunting and loss of natural habitats to the expansions of agriculture and logging. His concern with the latter problem has taken him to the Kruger National Park in South Africa, the Llanos in Colombia, the North and South Pantanal in Brazil and various zoos in Europe and North America. All this experience has given the author a profound respect for the entire feline family.

Preface

The overall well-being of the animals that are based in our society is gaining increasing amounts of attention. In response, the subject of animal welfare has developed into a major concern in the academic world. Animal welfare has now been incorporated in the study of veterinary medicine and other animal disciplines. The objective is to establish rational standards of animal care based on scientific observation, experience and basic empathy. A major, global problem in animal welfare pertains to the destruction of millions of unwanted or mistreated cats. Our concern for these animals can be improved by a better awareness and acceptance of their origins and essential natures.

Of all the domestic animals, the one whose nature has been least understood is the ordinary housecat. Even today, this principal pet is often subject to homelessness through failure to appreciate its nature and its inherited customs. In basic cat care there is much that is not in keeping with the fundamental nature of this animal. As a result there is a high displacement rate in urban cats. Animal shelters cope with great numbers of unwanted cats and kittens. In many cases it is the behaviour of these animals that is unappreciated and this is because of the misunderstanding of their natural ways. Many cats in animal shelters require to be euthanized and this constitutes a major issue in animal welfare.

Cats vary in their own individualities and attitudes, exhibiting a wide range of personality types. To appreciate the essence of cats requires some knowledge of their history. Archaeological findings show that this animal lived in some association with humans thousands of years ago in the Middle East. Specifically, they originated in North Africa, most notably Libya and, in particular, the Libyan Desert. After taking up residence in neighbouring Egypt, their natural dignity and comfort-seeking habits elevated their social standing to such an extent that they became deified by the households of Pharaohs. In addition, these cats became vital in the protection of the granaries that were forming the basis of Egypt's emerging civilization (Clutton-Brock, 1993).

The remains of cats from earlier times have been found in other places of human habitation. A few incidents of wild cats being tamed and kept in human custody do not create a new domestic species. The species of *Felis sylvestris catus* did not become established until there was a substantial body of tamed cats breeding together in perpetual human custody.

As Egyptian civilization developed about 5000 years ago, cats became widely popular due to their ability to catch rats and mice. As guardians of grain stores, they became invaluable. Many found themselves taken aboard ships to other parts of the world, adjusting easily to shipboard life and no doubt making lucrative cargo at distant ports. By such means, small populations of cats became established in widely separated regions, creating foundations for breeding stock in their new locations. They were then subject to human selection when variations in type emerged from mutation. Certain varieties got preferential selection based on popular promotion and local suitability, and subsequently, became breeds. Differing breeds of cats soon emerged in the isolation of these various locales. Even today, the development of new breeds still occurs in the hands of cat fanciers who aim for an interesting type, or mutation, to perpetuate.

All the many wild species in the cat family have so much in common with the domestic cat in their habits that it is compelling to review the lives of all wild cats. This illuminates the innate natures of common domestic cats that have become the principal companion animal in modern society. When the housecat is scratching, climbing, exploring, inspecting, stealing, sprinting, leaping or hiding, it is merely following instructions from its ancient genes in the same manner as the leopard and all the rest of the great feline family around the world. Information about their dynamic characteristics is needed to see and understand the full picture of this companion animal. Even the wild cats need to be better understood in this civilization since most of them are in dire circumstances and close to extinction. Wild cats serve a purpose in nature by keeping a balance in animal populations and controlling the species that can be a plague on human society.

Civilized society knows that it is inhumane to harm an animal deliberately and thereby cause it to suffer unnecessarily, so legislation decrees that it is unlawful to be cruel to animals. By human intuition, cats have been given the right to freedom from any harmful deed and the right to protection from injury and disease. Thus, society has given cats the right to proper care while in human custody. Such rights merit some expansion to accommodate improvement in the comprehension of feline welfare.

Cat keepers often see two faces, or dimensions, to their cats. There are times when their cats are mysterious. This is because, in reality, there are two sides to the domestic cat, the familiar side and an unfamiliar one, arising from the fact that the cat has a domesticated side and a wild side. The entire feline family has retained some ancient genes that go back for 10 million years. These genes dictate certain ways of living and the domestic cat has its share of such genes, so that this cat has certain instructions from its DNA that must be followed at times. Wild felines need better appreciation also in this time of human encroachment into their ancient lands causing them a major welfare crisis; they also have two sides to their existence, one is their home life and the other relates to human harassment. In addressing both aspects of the cat's nature it is sometimes necessary to use different terms, some in conventional style, some in technical style. A glossary of behavioural and technical terms is provided at the back of the book to explain them.

This book is intended for a readership of students committed to the provision of health and well-being to cats, and also for those who are involved professionally or vocationally with cat care. Feline management needs modern literature to turn to. Feline illness and disease are not a primary focus in this book; these are, of course, addressed in specialized veterinary texts (e.g. Little, 2012).

Acknowledgements

The author is grateful to Paul Layhausen who first aroused a special interest in feline behaviour in private discussion many years ago. Sarah Hulbert of CABI is thanked for her helpful encouragement to put the author's early notes into the form of a text, principally for veterinary students and feline devotees. Special gratitude is expressed to Christina Rees for her work in transcribing much of the text and contributing to its editing. Special thanks also go to Melanie MacDonald for her keen photographic work that supports the text throughout, and to Renee Harman, Denise Critch and John Woodman for their valuable contributions. Ian and Sean Fraser's help in tracking down literature is most appreciated.

Special acknowledgement

The author's elder daughter, Rhona Loptson BA, MSc, received her education in Scotland and Canada and now resides in Scotland. Her established interests include photography, animal welfare and fine art, specializing in animal subjects. With a qualification in mental health care, she has expanded her concepts in animal welfare.

When the final revision of this book became due the author was unable to attend to this, being scheduled for heart surgery. Rhona Loptson accepted the request and the challenge to perform the final revision, with all the detailed work involved. It will be understandable that the author is unable to express his gratitude adequately for such help, but it is profound.

1 Introduction

While the foundation of feline well-being is good health, a spirited manner of living is its principal identifying feature (Fogle, 2011) and is the focus of this book. As discussed in the Preface, manifestations of well-being are chiefly behavioural in form. All categories of behaviour that relate to the cat's maintenance as an individual reflect the animal's liveable state. Through the fullness of its behavioural output in maintenance, the cat's quality of existence becomes evident. To illustrate, normal activities exemplify a satisfactory state, but generous, wholesome activity reveals a quality. In order to recognize the variety of norms in general feline actions, it is necessary to appreciate inherent feline behaviour as it exists across the range of species that live throughout the world, from the massive Siberian tiger in the Russian Far East (Vaillant, 2010) to the miniature kodkod in the South American south-east (Sunquist and Sunquist, 2002) and very many other species of varying sizes (Table 1.1).

Developments in the fields of preventive veterinary medicine, applied animal behaviour and animal welfare allow well-being to become a feasible target in animals. One major feature of animal well-being, however, is that the subject must be treated as an individual, which is nevertheless a member of a species or society. It is easier to apply the principles of welfare to cats than to other domestic animals, since they are usually managed individually and their conditions and characteristic behaviours are familiar to those who care for them (Johnson-Bennett, 2011).

Cats have two simple goals in life: self-maintenance and reproduction. When neutered, life becomes even less complicated, leaving self-maintenance with its intrinsic pursuit of satisfaction as the cat's broad objective. In the *Shorter Oxford English Dictionary* of 1933, the word 'well-being' is defined as 'the state of being well, or in good health'. In recent times, the meaning has expanded in common use to refer to sound physical and psychological health, denoting a comprehensive state of wellness. This is the meaning that is used in this text in relation to our companion cats. It is taken that the minds of these animals are reflected in their behaviour. The complete picture of this behaviour is like an assembled jigsaw puzzle of the individual's sundry activities.

Over 2000 years ago, a healthy relationship between body and mind was acknowledged by the Romans in their dictum *mens sana in corpore sano*, a healthy mind in a healthy body. Perhaps theirs was the best definition of well-being. In any case, it is a very necessary term in today's concern for health management. 'Well-being' as a general lay term is often used as a synonym for welfare, but the latter is more of a concept that applies to groups of animals. The concept of well-being has been adopted in veterinary medicine as a fresh objective, occupied as it has been with exploring and defining disease, and developing specific and effective treatments to establish health.

Sentience

The general understanding and acceptance of the existence of animal sentience is astonishingly recent, in spite of centuries of documented evidence that animals' perception of their surroundings influences behaviour and affects their awareness and states of emotion (Fig. 1.1).

Sentience is a property of all higher animals, but welfare science is particularly attuned to it in companion animals, as they are viewed as fellow individuals. Compared to humans, they all possess well-developed senses and nervous systems that vary in degrees and types of perception. It can even be argued that domestic animals have a heightened sense of touch, hearing, smell, sight and taste. Through the property of sentience, animals experience pain and suffering, relief and pleasure. More

Table 1.1. Female cat sizes.

Average size	Species of cat
Very large (120–130 kg)	Lion, Bengal tiger, Siberian tiger
Large (35–65 kg)	Jaguar, cheetah, puma, leopard, snow leopard
Medium sized (8–16 kg)	Clouded leopard, Eurasian lynx, Canada lynx, Iberian lynx, bobcat, caracal, serval, African golden cat, fishing cat, jungle cat, ocelot
Small (3–5 kg)	Margay, oncilla, jaguarundi, Geoffroy's cat, pampas cat, European wildcat, African-Asian wildcat, domestic cat, manul
Very small (1–2.8 kg)	Leopard cat, marbled cat, sand cat, flat-headed cat, rusty-spotted cat, black-footed cat

Fig. 1.1. Mild fright.

than mere responsiveness to stimuli, sentience includes sensory evaluation of the various positives and negatives in their lives. Feline sentience includes the inner feelings associated with comfort-seeking, safety-seeking and, at times, thrill-seeking.

While sentience influences the response to stimuli, it is more psychological and emotional than physical and mechanical. Sentient animals experience suffering in various forms that can be acute, chronic, and nearly any other type of suffering that can be described in human terms. Acute pain, however, is the hallmark of severe suffering in both animals and humans, so any concept of animal suffering must include an appreciation of the impact of painful experience on the animal's well-being. The recognition of sentience in domesticated companions raises questions about the obligation incurred by society through such their use. It is argued that domestic animals deserve maximal humane consideration, because of their sentience and human moral duty. Welfare is now in the position to discharge this responsibility.

In very large measure, groundwork was affected by the growth of two relevant subjects, namely preventive veterinary medicine and applied animal behaviour. The publishers of the scientific journals on these two subjects deserve considerable credit for their foresight and bold decisions to treat these topics as respectable sciences. The body of objective information is presented for use in the perception of domestic animal behaviour and characteristics. Early seminal studies on the behaviour of felines, both domestic and wild, were made by Leyhausen and Falkena (1966), Leyhausen and Tonkin (1966), Leyhausen (1979, 1988) and West (1974, 1979).

The concealed state of mind in animals was first shed light upon by the Nobel Laureate of 1973, Konrad Lorenz. He is regarded as one the foremost naturalists of the past century, following the publication of his book *King Solomon's Ring* in 1952. The book's title relates to the legend that Solomon wore a magic ring that allowed him to talk with the animals. It was through their behaviour, however, that Lorenz's animals communicated with him. Cats, equivalently, have always expressed themselves through this means. By paying attention to this associate pet's behaviour, a proper understanding and relationship can be acquired.

The output of animal movement and action originates from the operation of the central nervous system, and of the brain in particular. The animal mind generates the thought processes that are basic to much of its behaviour. A cat's brain creates its behavioural dynamics, which indicates its state of well-being. Its show of sense, indicating intelligence, is of a type and level far lower than the human form due to a smaller cerebral cortex. Feelings, however, are largely processed within the brain's elaborate limbic system of both humans and felines. The cat's limbic system is highly functional, enriching its life and nature. Its expression of feelings reveals many of the emotional states commonly experienced in humans. Cats display their moods, wants, dispositions, temperaments and emotions. These are in addition to the widely acknowledged principal feline feelings, of pain, hunger, fear and rage. In the centre of the cat's neurohumoral system is its fixed character of predation. In the feline family, it is a means of survival, fully active in the wild cats and less active, but still present, in the domesticated species.

Without a doubt, cats show affection as a characteristic inherent in affiliation and it is clearly demonstrated in their domestic relationships through a physical, non-sexual fondness. This emotional link is typified by the parent–kitten attachment. Plato described this type of emotional bond as real love, and in line with this concept, many cats develop fond attachments to their caregivers. It is clear that some cats have an emotional capability, in response to the fond empathy expressed to them in their care (Frazier and Eckroate, 1990, 2008).

Conceptual Principles in Applied Feline Welfare

'Animal welfare' is the established umbrella term for the discipline relating to the protection of health and well-being in animals under human care. It further relates to measures of involvement in animal health. 'Applied feline welfare' is taken here to be the subject of practical, beneficial care for the cat's holistic well-being in domestication. In theoretical approaches to animal welfare, the two terms 'welfare' and 'well-being' are commonly used synonymously. The definitions of these terms in dictionaries allow either use, but their differential use is necessary here to distinguish accurately between what is extended exogenously through good care for the animal and what occurs in the animal's endogenous state of holistic soundness. In the terminology of this book, welfare and well-being are treated with their separate meanings. The main focus of this text is well-being, which is, in fact, the primary objective in welfare and preventive veterinary medicine.

In the practice of animal care, 'welfare', as a term, relates to the animal's supportive ambience while the term 'well-being' is identified as a health factor. An adequate provision of shelter and nutritious food, for example, are fundamental features of proper welfare for a cat; while good bodily condition and conative behaviour indicate a general state of well-being.

Among many external features of a good physical state, there would include a healthy coat, muscling on the back and thighs, bright eyes, good teeth and proper weight. As for psychological and emotional well-being, conative behaviour provides reliable evidence. Conation is willingness for action and performance of voluntary activity. Examples of conative cat behaviour are brisk kinetics such as spontaneous running and stationary exertions such as general stretching. The essential objective in applied feline welfare is therefore the establishment and maintenance of physical and mental well-being; the latter revealed through behaviour. The cat's natural behaviour is tuned to its survival (Poole, 1996).

With well-being as the target of this text, appreciation of its compound nature therefore requires a liberal approach. The physical aspect of this state can be addressed more objectively than the equally valid psychological aspect. The latter, in relation to animal subjects, can be managed scientifically through ethology, with certain affective states being identifiable behaviour. Within important aspects of welfare, it becomes necessary to accept certain posits about psychological well-being. These are logical and rational assumptions based on perceptions by a large body of cognizant observers, providing a structure for the subjects.

Feline Features

The various members of the cat family (Felidae), including the common housecat, have many physical and behavioural features in common. During movement, weight is taken on their toes and pedal pads. The forefeet have five toes while the hindfeet have four, carrying strong, curved, pointed retractable claws encased in sheaths from which they can be extruded by a specialized muscle called the flexor profundus perforans. An elastic ligament withdraws the claws.

The feline face is relatively short and the head has a rounded form with a compact muzzle (Fig. 1.2).

Fig. 1.2. Domestic cat face; note relatively small nose and large eyes, for close inspection.

The six incisor teeth, above and below, are small and arranged in a straight line. As a result, the large canines are placed close to the front of the mouth. There are three premolars on each jaw, but only one small molar per jaw. The condyles of the lower jaw are elongated and locked by ligaments in the glenoid cavities in a way that prevents any rotary movement in the temporo-mandibular joint. The tongue's surface is roughened and prickly due to the presence of horny papillae. The latter give the tongue a rasping capability for licking flesh from bone. The eyes are large, with vertical pupils capable of expanding to a circular opening or closing to a vertical slit.

These anatomical factors allow the cat to have a close sight on its feed, facilitating its fastidious nature in eating. The dentition is designed for tearing food rather than chewing it, propelled by strength in the jaw. The retina contains both rods and cones, allowing the cat to detect colour. The whiskers, or vibrissae, are sensors that assist in detecting the nature of items in front of the face. They are directed forward when the cat is hunting or being aggressive. The small species of cats have their visual sense tuned to close inspection, while the larger cats have their senses for odour and sight developed for long-range detection of prey. This difference is most marked when comparing the giant cats with the domestic species (Fig. 1.3).

Behaviourally, cats are light-footed and supple, capable of moving in a slow lurking fashion or suddenly sprinting at great speed for a short distance (Fig. 1.4).

Pure carnivores, cats hunt by creeping up on their prey before springing. Prey is immobilized by biting the neck region, either through choking or severing the upper cervical vertebrae in a pithing manner. The kill is eaten quickly under wild circumstances. With some wild species, an unfinished carcass may be covered over for eating at a later time. Cats are

Fig. 1.3. Contrasting faces.

Fig. 1.4. Feline walking style; note low carriage.

essential in nature for aiding the natural population to have balance in forms of animal life, such as rodents, that can proliferate to excess and be destructive to human resources.

Wild felines generally live solitary lives and are highly possessive of their territory. Hiding within this territory, they rest and sleep the majority of the time except when hunting or searching for a mate. Most wild cats are assisted in their secretive lives by camouflaged coats of mingled colours and remarkable patterns displaying a beautiful appearance. They mark their territory diligently by spraying urine on to prominent surfaces, depositing uncovered faeces at noticeable sites, or scratching tree trunks to make visible, vertical scars.

Dynamic indicators of well-being are the principal components of the categories of behaviour involved in the cat's occupation of self-maintenance, namely reaction, ingestion, body care, motion, rest, association, exploration and territoriality. These have already been described but some features require emphasizing because of their special relevance in the monitoring of well-being. They also deserve special appreciation as the foundation of the cat's character. This creature has always been undervalued as a companion animal and one reason for this may be the fact that while friendly, the cat has not been sufficiently sycophantic or subservient in its relationship with humans. Its independent, self-sufficient ways appear as a degree of detachment from the human scene. Its very firm, feline nature is there to be fully recognized, however, as an admirable quality for a companion (Thomas, 1994).

2 Well-being and Its Roots

With standard care and feeding, quality of life for domestic animals can sit anywhere along a scale ranging from survival to fitness. Well-being, on the other hand, requires a higher level of care that encompasses preventive medicine and enables healthy normal behaviour in each individual animal. Modern cat management often reaches that level and shows how the welfare of all cats can be improved. Better welfare means better living conditions and standards, for both the animals and their keepers. The end results of such policies are reduced suffering, improved well-being and better physical health in the animals that live as human companions. The ultimate outcome of humane animal care is a comprehensive state of health, which is the broad objective of animal welfare. This means good body condition or fitness, the expression of normal behaviour, signals of special health and clean, comfortable surroundings. In brief, the animal's behaviour and physicality reveal signs of well-being: when physique is normal and healthy, well-being is primarily on display through behaviour and must be sustained with appropriate care. Ten items of care are given in Table 2.1.

Sentience

Once an animal's nature has been keenly observed by its caregiver, it remains apparent and cannot ever become unnoticed. When the richness of a cat's self, with its sensitive properties and firm characteristics, becomes fully recognized by its caregiver, it will then become appreciated as a being of quality. This appreciation is the best basis for a good animal care policy. Well-being in a cat is mainly dependent on its own emotional balance and good routines of care (Davidson, 2011).

Affective Factors

Emotive quality

Contrasting qualities of feeling are evident in a cat's display of anger, affection, dislike and fright. Common actions, such as scratching, biting, hissing, ear flattening, hair raising and calling, indicate a variety of brief feelings. More durable feelings exist as moods (Fig. 2.1).

The extreme moods of joy and depression can be identified in such behaviour as saltation and unresponsiveness, respectively. Recurring saltations are apparently euphoric displays, while undirected wandering and restlessness can be symptomatic of dysphoria. Euphoria in the cat is the expression of well-being in such acts as lively kinetics. These expressions may be in the form of bolting through rooms or climbing curtains. Dysphoria, however, is an emotive property indicating an enduring state of mental discomfort expressed through loss of appetite, low-grade activities and soiling. When such conditions become habits in the cat's behaviour, they amount to a psychological disorder.

In their relationships with people unknown to them, many housecats exhibit one of two possible social attitudes. They may display a positive attitude, by taking an interest in association by approaching the individual readily and remaining in close proximity. Further, cats with this sociophilic attitude are consistent with friendly behaviour, tending to investigate any new arrival in their home territory. By contrast, some housecats consistently show a sociophobic attitude towards strangers. These cats hurry away from the individual, keeping their distance without attempting any investigational approaches. This avoidance is usually maintained for hours before any tentative move to enter the newcomer's presence. In an extreme form, a sociophobic cat will remain in hiding for extended periods after being alarmed.

Table 2.1. Outline of routine cat care.

Item	Method	Schedule	Comments
1. Tending litter	Scooping/changing	Daily	Note normal content
2. Feeding	(i) Wet	(i) Twice daily	a.m. and p.m.
	(ii) Dry	(ii) *Ad libitum*	
3. Watering	Renewal	Daily	Fresh, not flat
4. Grooming	Brushing	Daily	Overall, including underparts
5. Smoothing	Manual	Frequent	Wipe face, head, back
6. Ledges	Clearing	Daily	Upper levels for rest
7. Inspection	Visual checking	Daily	General monitoring
8. Play	Available toys	Regularly	To induce activity
9. Sleeping quarters	Tidy	Daily	For comfort
10. Lifting	Raise and carry	Occasional	Handling, assess weight

Fig. 2.1. Anger (top) and attention (bottom).

Sociophilia and sociophobia are not characteristic features in the general attitudes of cats, and do not appear evident in their behaviour during any other time. In addition, these attitudinal traits do not appear to have any relationship with upbringing or prior experience, and can be categorized as inborn. The sociophobic cat may not show unusual fearfulness in various other circumstances, and the sociophilic cat may not be particularly friendly towards its caregiver in its general daily conduct. These traits appear to be inherited and the related genes may be widely diffused in the common cat population, as it appears that most cats are likely to show one or the other of these characteristics. A minority of cats, however, display modified attitudes to unknown individuals.

Mood durability

In many cats, certain moods are chronic in nature and typify their individual temperaments. Varieties of characteristic temperaments in cats are easily recognized by their owners. Such temperaments include such subjectively termed conditions as compliance, flightiness and aggressiveness. Additional feline temperaments are recognizable according to the customary disposition of the individual animal. These are means of profiling the subject's general, typical attitude.

In some cases, a cat's temperament can be altered by experience or training. Timidity and nervousness, for example, can be adjusted by accustoming the animal to a range of common circumstances through repetition of non-aversive experiences. In other cases, temperamental features may not be changed. An aggressive cat will usually retain its temperament, despite attempts to remove it through training. As a general rule, feline temperaments are durable and may have a basis in kittenhood.

Brief states of mood are very commonplace in feline behaviour. To illustrate, cats are often quite excited during food preparation and calm after feeding. With the arrival of a strange person, some housecats become temporarily hyperactive or suppressed, as discussed above. Young kittens are usually distressed following separation from their mother, but are soon pacified when reunited with her. Many cats become agitated if their morning feed is delayed, but relax immediately upon feeding. In these various instances, the animal's display of mood can be taken as an expression of feeling. A state of depression is frequently associated with feline disease, in which the animal is incapacitated.

Persistent disposition

Some studies on the social behaviour of cat groups have been aimed at determining rank order, on the prior assumption that such a fixed order occurs in this situation. While group individuals may be assertors or evaders, such tactical differences contribute to group peace through assuagement. In addition, the formation of bonds within an established group is usual and contributes further to social cohesion. Clear rank orders are not normally featured in the group behaviour of this animal (Crowell-Davis *et al.*, 2004). When disruption occurs, however, it is not a function of rank, and acts of fighting are not always significant. A generalized disposition of passivity thus typifies the behaviour within an established, closed group of ferals. This disposition, however, does not relieve individuals of their own, given individuality. The latter is more apparent in human custody.

Sensory poles

In the matter of human feeling, the maxim is that pleasure is the antithesis of pain. Like people, cats can show clear signs of pain, as evident in serious trauma, for example. It is certain they experience true pleasure, as apparent through a deep, throaty purring in situations where they appear at ease and seemingly satisfied with their immediate circumstances. Differences between the opposing sensory states of pain and pleasure in the cat are given below:

- Pain has a neurogenic transmission, through substance P, while pleasure is an affective state resulting from dopaminergic effects.
- The experience of pain is instantaneous; pleasure is a less immediate experience.
- Causes of pain are identifiable, while causes of pleasure in a cat's common circumstances are not always obvious.
- Pain can be caused quite easily, while pleasure cannot.

In order to assume that the cat experiences pleasure, it is necessary to employ anthropomorphism, since proof of its existence is only in supposition. Anthropomorphism is an old liberal concept that, uncritically, attributed to animals such features that are typically or essentially human. In conventional science, such an attribution has been deplored, based in turn on the anthropocentric concept that animals are not human and humans are not animal. The latter view is presently not held so firmly outside of religion. A more common acknowledgement today, largely due to Darwin, is that the human is an extraordinary species (Jones, 2001). An objective comparison between the cat and the human in major factors, such as structural and organic anatomy and functional physiology, indicates that the two species are analogues of each other. The cat's front feet, for example, are equipped with five digits and a rotating wrist – equivalent to human hands, with claws analogous to fingernails. Like the human head, the cat's head contains a brain, mouth, nostrils, face, jaw and sense organs. The reproductive organs are also analogous to those of a human. Such analogizing can continue endlessly, indicating that resemblances between the two match their differences.

It is a logical assumption that the cat has an affective constitution that is to some extent an analogue of the spectrum of human emotionality. Such use of analogy is not blind anthropomorphism, but a rational method of appreciating the feelings that the cat could experience covertly, in accord with its welfare. It may be that where a negative emotion is evident in the cat, a positive contrast is likely to occur. A feeling of discomfort, for example, can be displayed in behaviour; but comfort, as contrast, could be experienced. Fear can be exhibited, and by contrast, security can likely be felt.

Other contrasts of feeling that can be ascribed to the cat include: hunger and satiety; hyper-excitement and placidity; rage and peacefulness; loneliness and companionship; exhaustion and restfulness; hostility and friendship. In addition to polarities of feeling, median states are also experienced. The negative poles of these affective states are more evident expressions of feeling through exaggerated behavioural displays, while discrete signs of positive feeling are recognizable through focused study. To illustrate, in positive states of feeling there is purring, and vocal sounds are minute. The tail is carried upright and the ears are held forward, indicating the animal's approachability and acceptance of handling. In support of an acknowledgement that a comparative system of affect exists in the cat is the fact that the cat brain possesses a limbic system comparable to the human limbic system, as mentioned previously.

The analogical view is limited to basic similarities; where a fundamental difference exists, divergence occurs. While the human possesses cognitive ability, the cat does not have such a comparative property. Some neuroplasticity can, however, create fresh neural circuits for abnormal or novel activities such as cloth chewing or living compatibly with an alien species. Similar neural biochemistry does however exist in both species. Humans and cats have limbic systems with neural biochemistry, synthesizing emotions. They do not share a similar prefrontal cortex, that which provides a high level of cognition. On their cortical substance, analogization breaks down. Divergent differences are massive between both species. The cat's great mutational gifts were the acquisition of a small body, sharp canine teeth and retractable claws, while human mutations included a massive cerebral cortex, an upright stance and manipulative hands.

Positions on suffering

Most commonly, cats experience emotive suffering in addition to disease or injury. In such cases the prominent elements are pain or malaise. These elements are usually overtly expressed through behavioural changes. Similarly, changes in the cat's behaviour can occur when suffering is the result of neglect or abuse, and as a rule, they are not so overt or prominent in nature. A condition of aboulia, or loss of positive will-power, typifies feline suffering that lacks physio-pathological basis. The condition of learned helplessness is an example. Cats in such a state lend themselves to human cruelty.

Suffering and well-being are opposite states, either tolerated or enjoyed by the overall constitution. It is a valid clinico-behavioural rule that the development of either state displaces the other. The imbalance between these two endogenous states demands the recognition of both in behavioural manifestations. Suffering is a component of amorphous conditions such as stress or distress that pertain to the psychological state of the subject. The nature of anguish in animals is intangible, although substantial subjective evidence attests to its existence. The clinical arrest of normal behaviour is assumed to be linked with suffering, and recognition of this unhealthy state requires appropriate experience and medical knowledge of behaviour and health. In clinical veterinary work, levels of suffering are implicitly understood in such clinico-behavioural manifestations as collapse, arresting pain, passively depressed behaviour, anorexia and significant inactivity, in an order of diminishing degrees of disturbance.

Suffering is also a component of more ambiguous conditions such as stress or distress. Distressful circumstances of custody can disturb the cat beyond its ability to adapt to them. Suffering can therefore result from psychological insults, independently of a physical cause. Certain acute behavioural signs such as intensive vocalizations, struggling and trembling are clear evidence of a reactive variety of suffering (Fraser and Quine, 1989).

A simple equation can illustrate the concept of animal well-being, as follows:

Normal Health + Natural Behaviour
= Animal Well-being.

The state of being comprehensively well has the support of three principal theories.

1. The animal's health and existence can be optimal by accommodating the main features of its natural behaviour including free movement, and by providing quality care, as far as is practicable in each case.
2. The guidelines in animal management and use to support well-being must be based in bioethical principles that are favourable to the animal.
3. In animal well-being, the main contributing factors are quality of husbandry, protective health care and stress control; stress being regarded as a significant neurochemical disturbance in the animal.

The cat's behaviour can be regarded variously as free-style, fixed-style or mixed-style, as stated below.

1. *Free-style behaviour* demonstrates the animal's motivation, interest, thinking, decisions and inherent capability. Its learning and capacity to interact with its surroundings is considered free-style. Much of our cat's daily output of behaviour falls in this category.
2. *Fixed-style actions* are typically thought-free and inflexible. These stem from what can be regarded as firm 'wiring' (neural centres and pathways in the brain), which functions in response to fast-shunting chemistry. In these instances, such chemicals as hormones and neurotransmitters are induced in response to a triggering factor. Firm neurochemical arrangements generate fear or anger, fight or flight. Strong instincts fall into this category; they can be conveniently regarded as inherited units of behaviour.
3. *Mixed-style behaviour* is emitted by established 'wiring' and compound chemistry affecting the brain. Knowledge through learning can modify natural behaviour inherited from the animal's ancient ancestry. This is also partially in the form of instinctive cat activity, as is observed in its reproductive behaviour.

Philosophically, it can be said that people and cats have one major interest in common: the motivation to achieve and retain well-being. The appraisal of animal well-being is based on both the dynamics and the physical appearance of the animal by assessment of the particular individual. The physical features of relevance include the coat, eyes, flesh and bones. The coat, from head to tail, should lie naturally with its normal direction of flow and should not appear dry, harsh, dishevelled, tangled or soiled. The eyes should not be sunken or fast moving, but exhibit a sparkle or glistening appearance. The physical features of the entire body are of primary importance. A degree of fleshiness should be detectable over the back and protruding ribs should not be noticeable. The thighs should be firm and muscular, as is appropriate for the feline's age and species type. A healthy cat can be observed to walk casually.

The cat's dynamic features of well-being are furthermore quite specific and include eight categories of behaviour. In general, these are the classes of behaviour involved in the self-maintenance of the animal, as follows: (i) reaction; (ii) ingestion; (iii) body care; (iv) motion; (v) territorial activity; (vi) association; (vii) exploration; and (viii) rest. The details of normal and beneficial behaviour in these categories will be fully addressed later in other chapters. At this point, however, it is appropriate to draw attention to the fact that cat care must address the needs involved in the animal's self-maintenance behaviour for the purposes of establishing and preserving well-being.

Professor Bernard Rollin, who holds positions in philosophy, biophysics and bioethical planning at Colorado State University, is the world's leading authority on the ethics of animal use. He has repeatedly pointed out, in his numerous publications concerning the treatment of animals, that humans alone are moral beings. Rollin (1995) indicates that civilized people require a moral position in animal use. This demands a moral trade-off through the exchange of their lives and companionship for a humane level of animal care. Cats are psychological human auxiliaries. Unfortunately, the necessary destruction of countless uncared for and unwanted cats by animal control authorities is a condemnation of our morality. The international scale of this annual destruction of displaced cats is in the millions.

No evidence exists to show the dependence of emotionality on intellect. A person does not have to be clever to enjoy, or to suffer. The lower level of intellect in cats in no way precludes their ability to suffer emotionally as well as physically from an illness or a traumatic experience. Their general sentience is sufficient to make each individual cat aware of, and susceptible to, violation to any part of its own comprehensive condition, namely its well-being.

In multiple-cat households, owners can recognize the emotions of their felines through episodes of jealousy between the individual cats when one is ousted. Generally, cats are suspiciously vigilant of each other. Resident cats can experience lasting resentment towards an introduced cat on account of suffering from sudden rivalry. This is most likely to occur when three cats are present in one household. When left alone, many cats show a disaffected attitude towards their caregivers on their return from a protracted absence.

Components of well-being are required to coexist and to be in harmony. The animal's accommodation for the definitive totality of this state is of high importance. Animal well-being is therefore a union of factors. Its components are physical health, dynamic capability, sensory satisfaction, ingestive sustenance, body comfort and appropriate accommodation. Proper animal care establishes all these things; in particular, the recognition of animal well-being as evident health and manifested contentment, as is typically observed in replete and resting litters of kittens. When such a litter is seen, this image of well-being can provide a lasting reference. Small, full bellies, the dozing assembly, the absence of restlessness, litter mates bedded together against the maternal abdomen and sheltered between the mother's fore and hind legs; all of these depict well-being.

A knowledgeable check on litter health would be required, as the kittens could have parasites. In such cases, the administration of anti-parasite treatment would ensure their well-being. The health check would also include an overall inspection for any abnormalities, including genetic defects. Any untreatable congenital condition that would result in later suffering would, however, call for prompt euthanasia (welfare has a preventive side). In addition, the health check would extend to the mother chiefly to determine a good supply of milk, but also to ascertain the presence of any communicable disorder. On the topic of prevention, it is well to realize that preventive care is economical and easier than treatment for a clinical disorder such as a preventable disease. Well-being cannot coexist with illness.

Sensors for Well-being

Hearing

Cats have a keen sense of hearing (sharper than canine hearing) and are able to detect high-pitched sounds that are ultrasonic to the human ear – up to 65 kHz in range. Loud noises such as blaring radios, television sets and vacuum cleaners are therefore hurtful to a cat's ears. An exception is white cats with blue eyes, which are genetically deaf from birth. Another well-developed sense is the cat's ear system controlling balance, allowing the cat to walk along narrow ledges and rotate its body quickly in a fall so as to land in a balanced posture, absorbing impact. In addition, their specialized sense of balance prevents motion sickness during transportation by car if they can alter their position with ease.

Smell, taste, touch and sight

The large cats have their sense organs for sight and smell designed for distance detection, their eyes are comparatively small and their noses are large. By contrast, the much smaller domestic cat has sensors for close detection, their eyes are comparatively large and the nose is small and fine.

While cats cannot see in total darkness, they have great visual ability in dim lighting – far better than the human eye. The cat's eyeballs are large relative to the size of their bodies. The inner structure of this eyeball has a mirror-like section on the rear retina which can spread out light coming through the pupil, allowing the animal to detect objects in the dark. When dilated, the cat's vertical pupil is circular and occupies nearly all the exposed eyeball. The lens is permanently set back providing great focusing ability and allowing for the rapid adjustment of changes in distance. The cat has extraordinary eyesight that includes the detection of colour. By way of contrast to this attribute, kittens are born with their eyes closed for the first week of life and do not develop their full vision until they are about 3 weeks of age.

As with many other animals, the cat is receptive to photoperiod. This is the slightly altering duration of daylight in each 24 h period, which works in contrast with darkness to create a stimulus via optic tracts to the brain's pineal gland. The pineal produces the enzyme hydroxyindole-*o*-methyltransferase (HIOMT) in response to the appropriate photoperiod. HIOMT is the precursor to the pineal gland's hormone melatonin, which in turn stimulates the hypothalamus to send a releasing hormone to the pituitary gland. The pituitary puts the gonad-stimulating hormones into the general circulation and breeding capability is established in any seasonally breeding cat.

The complexity of the above system indicates vulnerability at its various links, should stressful events alter the brain's chemistry. The importance of photoperiodism in the cat's environment is evident in its reproduction. Cats breed most readily when light quantity increases over the amount of darkness in each 24 h period. As a result of these environmental characteristics, the natural breeding season of wild cats in latitudes outside the tropics is early spring and summer. The Siberian tiger in northern Asia, for example, breeds in early spring, while the tropical Bengal tiger breeds at any time of the year. All tropical wild cats breed perennially.

The various senses in cats work in corporation with each other. Cats will often smell their food closely and then touch it lightly with their nose-pad before sampling. If this is satisfactory, the animal will commence eating. Sense of smell also plays a role in territorial actions. In addition to physical mark-making through clawing, cats spray urine on their sign posts, signalling their odour to other cats who will then take suitable action according to their interest in the sign. A sharp sense of smell can also be noted through their delight in the aroma of catnip. The sexes, if intact, contain odorous messages that convey meaning to one other in the business of reproduction.

The main system for odour detection is the nerve supply from the nasal passages to the olfactory area of the cerebral cortex, via the brain's olfactory bulb. Like other domestic animals, the cat has an accessory olfactory system called the vomeronasal organ (not present in humans). This consists of a pair of minute, sensory tubes located in the nasal passages for specialized detection of fine odour. The tubes have a connection with small apertures on the palette between the upper corner incisor teeth and the canines. Odour is inhaled by sniffing, with the mouth slightly open. The organ's main function is to detect breeding scents.

Taste is another major sensory factor in guarding well-being. It may function as a component in the cat's propensity to lick a chosen associate, and likely plays a part in a mother cat's recognition of her own kittens, after she has cleaned them at birth.

The sense of taste is important in the selection of appropriate food. Like smell, taste is the result of interactions between chemical stimuli and receptors located on the surface of a mucous membrane. In this case the receptors are taste buds on the tongue. Olfactory cells in the nasal membranes can combine the sense of smell and taste, explaining why both senses must be considered together. Chemical senses, both smell and taste are dependent on contact with chemical substances for stimulation. This is in contrast with the other senses that are dependent upon energies of light, vibration or pressure.

The taste buds are end organs in the mucous membranes of the mouth. The majority of them are located on the anterior two-thirds of the tongue. In taste, monosodium glutamate appears to facilitate the binding of chemical stimuli with molecules on the membrane of the tongue over the taste receptor. Molecules are transduced into a gustatory impulse conveyed to the brain and to a level of consciousness. Some differences occur in taste detection and preference for items of food. Such preferences involve conscious selection of items for ingestion.

Like smell and taste, touch is also an all-important sense in cat life. The paws are tactile receptors and the nose-pad is a touch organ with various uses, such as for checking the warmth of neonatal kittens. The cat's long whiskers and eyebrow hairs are well known as keen touch receptors, but the numerous, lengthy guard hairs covering the body also play a role in this business. Cats respond positively or negatively to their tactile experience; such is the importance of this sense in their well-being. The long hairs on the muzzle and brows are called vibrissae and are the most acute touch sensors. The whiskers, termed mystacials, are quite mobile and are also used as signals of disposition or intention, directed forward in aggression and backward in fear. Forward-directed whiskers in a fight indicate that the cat, when it contacts the foe, intends to bite seriously. Mystacials, when turned back, are streamlined for flight.

Awareness

When the cat is awake it maintains a high degree of sensory consciousness, as indicated by its typical prompt responsiveness to minor stimuli in its immediate environment. The most frequent response is alert, focused attention directed at the stimulating circumstance. This reveals an acute state of awareness that extends into exploratory behaviour. The mental energy expended in the maintenance of awareness at a high level is a demand on the cat's neurophysiological resources. The reserves can only be adequately restored in the unconscious condition of sleep and in such episodes that provide a considerable accumulation of sleep daily. This requirement is the cost of the cat's percipience – its high level of awareness at wakeful times. Such special awareness is a part of the compound suite of characteristics needed for this animal's style of predation on rodents, birds and fish.

Humans and sensory developed animals, such as cats, share the hippocampus – a major part of the limbic system. The hippocampus consists of elongated ridges on the floor of each lateral ventricle of the brain, and is considered to be the centre of emotional senses. Cats, therefore, have the capacity to experience emotions. This would only be expected in a sentient animal, with sentience having a cerebral foundation for awareness.

Tactile sensation and pain

Each bodily sensation experienced by the cat originates in a receptor and receptors are of specific types for the detection of particular stimuli. The receptor is composed of a nerve ending and a non-neural cell, specific for the detection of stimulation from pressure, touch, heat, cold, joint movement or pain. Pain receptors are located at the ends of small neurons, while the receptors for pressure are called the Pacinian corpuscles and are found in the skin. Somatic reception comes from mechanical forces acting on the coat, skin, subcutaneous tissues and joints. From the receptor, action potential frequency is established in the nerve and is transmitted to the central nervous system for appraisal. The stimulation of pain results in the release of pain mediators such as bradykinin and prostaglandin, which act on specific receptors of pain (Soulsby and Morton, 2001).

On stimulation of pain receptors there is electrical potential, where the frequency of the latter increases in proportion to the severity of stimulation. Such severity can be created by trauma caused by tissue damage, irritation or abrasion. Specific neural pathways for pain reception proceed to pain centres in the central nervous system; the cerebral cortex then establishes the conscious perception of pain strength and position. Other non-specific neural pathways of pain move to the brain stem and the limbic system for responsiveness. The brain stem creates reaction, while the limbic system organizes the autonomic response. Behavioural indicators of pain in cats are given in Table 2.2.

Sympathetic effects are evident in any of the emotions of fear, anger, fight or flight contained in the physical reaction. Extreme parasympathetic effects are evident in shock resulting from major trauma. Such effects include collapse, redirection of blood volume to the gut, a significant drop in peripheral blood pressure, reduced consciousness, thermal loss and possible wounds and fractures. These circumstances can often be found in cats that have been severely injured in traffic accidents. Death can ensue if the systemic effects are not reversed promptly. Twelve behavioural indicators of well-being, or wellness, are given in Table 2.3.

Pedigree Breeds and Their Characteristic Domestic Behaviour

The development of pedigreed breeds of cats has had the effect of giving each member of each breed similar forms of breed temperament that facilitate their domesticity. This invites a higher quality of care that enhances their well-being. Even among and within the purebred categories of domestic cats, the rule still applies that no two cats have identical natures. However, with the development of the pedigreed breeds that now exist, there has been great effort by the breeders to establish characteristics, both morphological and behavioural, which are typical of each breed. In this objective, remarkable success has been achieved regarding the special morphological features that identify each particular breed. With many behavioural modes being related to types of physique and coloration in the cat, some standards of common behavioural traits in each breed have been obtained, as described below (Edney and Bessant, 2001; Hackett, 2012).

Table 2.2. Behavioural indicators of pain in cats.

Seat of pain	Miscellaneous indicators/signs
Head	Droopy eyelids, fixed staring, immobile carriage of head, pressing head against upright surface
Ear	Tilting head, pawing ear repeatedly, shaking ears
Eye	Rubbing eye, watering from part-closed eye
Mouth	Drooling from mouth, reluctance to eat, scraping mouth
Throat/stomach	Retching, head and neck extended, drooling, coughing, difficulty swallowing, vomiting (hairball or food), howling
Abdomen	Turning to look at abdomen, licking an abdominal site frequently, sitting hunched up, immobile, panting, resists handling, vomiting, diarrhoea, constipation, refusal to eat, howling, growling, hiding
Rear end	Turning to look at tail, 'scooting' along floor
Structural parts (orthopaedic)	Loss of free movement, abnormal carriage, arrest of full pandiculation, difficulty in jumping, stiff movement, lameness

Table 2.3. Twelve behavioural indicators of cat well-being.

Postulated order of significance	Indicator	Incidence/occurrence
1	Spells of play, bursts of chasing, racing and/or climbing	Daily, especially in first three age groups
2	Episodes of stretching	Regularly
3	Yawning	After periods of sleep
4	Grooming in all forms, including washing face, temples, ears	Several times daily
5	Wiping side of head, arched body and hip against caregiver	Periodically
6	Responsive tolerance and posturing	Usually when stroked
7	Purring at length	When petted
8	Vocalizing gently/silent mewing oral action	On greeting in mornings
9	Obstructing caregiver's footsteps	Occasionally
10	Reflex tail erection	When stroked firmly from croup to tail root when standing
11	Watchful perching/crouching on elevated sites	Daily on a few occasions and locations
12	Licking associate cat or keeper	Occasionally

Table 2.4 gives a list of many pedigreed breeds with their geographical origins.

The primary breeders of the pedigreed breeds attempted to ensure that the behaviour of the breed had some uniformity, especially with regard to desirable features of human association. One principal object has been to make them into good pets. Here again, great success has been achieved to the point that each breed has its typically attractive appearance and suitable manners, given an occasional atypical individual. In a range of 25 of these breeds, a very brief account is given below of the typical domestic behaviour occurring in each, with particular relevance to habits in the domestic setting, such as activity level, 'talkativeness', association with caregivers and degree of friendliness. These features add up to the feline personality which, in its turn, may be mainly introverted, extroverted or some variety of either or both. The talkative ones are the most extroverted, as a rule.

Abyssinian

This breed is characterized by talkativeness and its style of athletic activities. It is energetic in its routine movements around the house, being busy and curious about any domestic activities taking place near it. This cat is very communicative with persons it knows and has a friendly disposition, although it may be indifferent towards strangers. Its energetic nature and its independent manners do not qualify it as a lap cat.

Table 2.4. Cat breeds and their geographical origins.

Breed	Country of origin
Abyssinian	Egypt (Ancient)
Balinese	Burma
Birman	Siam (Thailand)
Bombay	Burma and England
British Blue	England
British Shorthair	England
Burmese	Burma
Burmilla	Burma and Persia (Iraq)
Chocolate Longhair	Turkey
Chocolate Oriental Spotted	Siam (Thailand)
Cornish Rex	England
Devon Rex	England
Exotic Shorthair	England and Persia (Iraq)
Japanese Bobtail	Japan
Maine Coon	New England, USA
Manx	Isle of Man (England)
Oriental White	Siam (Thailand)
Persian	Persia (Iraq)
Ragdoll	Siam (Thailand)
Russian Blue	Russia
Scottish Fold	Scotland
Siamese	Siam (Thailand)
Somali	Egypt (Ancient)
Turkish Van	Turkey

American Shorthair

This breed has an agreeable personality. It has an affectionate nature towards its caregivers and has a relaxed manner in its affairs about the house. It is not overactive in its general behaviour but is definitely not a shy cat. Its vocalizations are quiet, yet it purrs loudly and readily. It enjoys being held as a lap cat and is gentle with children. It generally enjoys contact with visiting strangers. It readily accepts other pets, including dogs, as welcome associates, especially in the absence of the owner. It is a good hunter.

Balinese

The Balinese is a variant of the Siamese breed. It is very sweet-natured, intelligent and talkative. It is energetic in its household activities and is always keen to be involved with domestic affairs. It has a friendly nature and is sociable enough to accept a bit of lap time with the persons known to it. It enjoys spending time with the people of its household. This includes grooming sessions. Some owners remark that this cat 'sticks to me like glue' and comes when called.

Bengal

With Asian leopard cat blood in it, this cat is physically active by nature. It shows great curiosity in events in its environment, to the point of being entertaining. It has a particular affinity with water.

British Shorthair

This cat has a very warm disposition and an agreeable temperament. It is a friendly cat while being very independent in its household ways. It is therefore not demanding of attention and is not vocal while it associates with its keepers. Since it is self-sufficient, it can take care of itself and is a reliable pet. The belief is that it is not too far removed from the broad population of non-pedigreed cats in Britain.

Burmese

This oriental breed is extremely extroverted and affectionate, devoting itself to its people. Sometimes described as a cat for the connoisseur, it is very responsive to the prevailing mood of the household. By nature, it is very talkative with subdued vocals, usually directed to its human caregivers. It is very demanding of attention with evidence of a strong ego and does not like being left alone. Naturally, it does not appreciate being disregarded in the presence of its people. Although it is a strong, muscular animal, it is playful and also has its sessions of being a relaxed lap cat. This cat dislikes any other cat intruding into its domain, showing instant aggression to it and fighting it in a formidable fashion. When it is outdoors, it can be a ruffian.

Cornish Rex and Devon Rex

Given that these two breeds have close geographical origins, it is not surprising that they have an abundance of features in common. They are very active and talkative cats and have a high level of curiosity in their domestic affairs. Their intelligence leads them into many empirical activities that mark them as being mischievous. The Devon Rex is more of a lap-sitter than its cousin from Cornwall. Both have the habit of occasionally wagging their tails in the canine fashion.

Egyptian Mau

This breed is very affectionate and particularly enjoys lap-sitting with persons known to it. It is also very 'chatty' and can be very active around its premises. (The word 'mau' means 'cat' in Egypt.)

Himalayan

Above all else, the pedigreed Himalayan is a lap cat. Its need for attention and approval from its caregivers is so severe that this cat can pine for attention. It is not servile in its attention-seeking, but very expectant. It is very talkative and can sometimes engage its caregiver in conversation, expecting responses to its vocalizations.

Korat

This cat has an extremely close bond with its guardian and demands attention. It does not mix well with other cats, but responds very well to all persons in its household and likes to participate with them in various activities. Naturally, it is fond of lap-sitting, in spite of having a high energy level at other times. It loves to intrude and obstruct its owner when the latter is occupied in any activity that results in the cat being overlooked.

Maine Coon

This is an old breed of large cat brought by pioneers to the north-east of the USA, but it is now of pedigreed status with fixed characteristics. It is a very friendly animal and is talkative with a soft voice. It can appear somewhat dog-like in its manner of following around after its caregivers, although it is not a particularly active cat. It loves water and is content to walk satisfactorily on a leash with its owner. It is fond of children and dogs.

Manx

Cats of the Manx breed are known to be quiet and affectionate animals with a long history of being good household pets. They are typically 'one person' cats in the way that they select one individual to follow and talk to. They call for attention with a variety of chips. Being firm egoists, they do not enjoy the presence of other cats.

Norwegian Forest Cat

This cat is of ancient heritage and, as a pedigreed animal now, is very fixed in its behaviourisms. It is active, affectionate and talkative but resists being petted excessively. It is not a lap-sitter (Fig. 2.2).

Ocelot

Starting as a natural wild cat in northern regions of South America, this cat has now become accepted as a breed of domestic cat in some jurisdictions, not to be confused with the unrelated ocicat. It is now recognized as an exotic cat, requiring very specialized and sensitive care since it has a defensive disposition and is fearful of many ordinary domestic events involving it. As it ages, it usually becomes quite friendly with its principal caregiver and enjoys being involved in forms of rough play with that person.

Fig. 2.2. Norwegian Forest Cat.

Ocicat

This is an extrovert breed of cat with confident ways. As such, it likes the company of people and also of dogs and cats. It takes readily to being walked on a leash and has a high level of energy output. It is not particularly talkative or inclined to lap-sitting, but can accept sessions of the latter quite comfortably.

Persian

This most popular cat is appreciated for its gentle ways and characteristic calmness. Although it has a most benign nature, it can also be very out-going. It commits itself to its people as shown in its devotedness to them. It is not a talker, but is an enthusiastic lap-sitter. Its output of energy in its domestic activities is average.

Ragdoll

As its name implies, this breed of cat is extremely languorous. It is not very active or vocal, but enjoys being held on a lap and being treated like a doll. These characteristics make it a very good family pet.

Russian Blue

This breed is not overly affectionate and is not a lap cat, being dignified by nature. It is graceful in its movements and its activity level is modest, which gives it a style of being consistently calm.

Scottish Fold

Here is a true lap cat. It is very affectionate and has a good attitude with children and dogs. It can be demanding of attention, but is not a talker. The level of its general activity is quite average.

Selkirk Rex

Since it is a large, strong cat it is not built for lap-sitting, but it is happy with human company. It is not particularly talkative. Its activity level is quite high as would be expected, considering its physique.

Siamese

Of all cats, this may be the most self-confident and certainly the most talkative breed. Its vocalizations are very varied in pitch and tone, giving it the appearance of trying to communicate specific thoughts. Some Siamese owners learn their meanings. It is a cat that creates strong bonds with certain people it knows. It demonstrates the existence and strength of the feline ego. It displays its intelligence in a variety of ways of novelty in its behaviour.

Siberian

Cats of this breed are considered to be occasionally 'chatty' with their keepers, but not talkative at other times. They are certainly intelligent enough to have this trait. They are friendly enough to participate in lap-sitting when they are taken up. The level of activity in these cats is average (Fig. 2.3). The Norwegian and the Siberian are of similar appearance.

Fig. 2.3. Siberian cat; note resemblance to Norwegian.

Tonkinese

Although of medium size, this cat is strongly built and has a high level of energy, but manages to be a good lap cat. It is remarkably talkative in that it can have conversations with people by responding vocally with its keepers when spoken to in an agreeable manner.

Turkish Van

If there is such a cat as an all-rounder, this may be it. It is an extremely active cat that is also very affectionate and fond of lap-sitting. It is very talkative with its keepers and has a balanced ego.

Unknown Senses

It has long been generally accepted that some cats can possess a sense of certain environmental factors that are beyond human detection. An irrefutable example of one such 'extra' sense in a cat is given in detailed case histories by Dr David Dosa, a geriatrician and Assistant Professor of Medicine. He describes the phenomenon in his factual book *Making Rounds With Oscar* (2010). One particular cat had the routine habit of affiliating closely with any patient about to die in a large nursing home in Rhode Island, USA. The imminent deaths were indicated by the cat's connection to the patient before the medical staff, or Dr Dosa, knew of the moribund developments.

Conclusion

It is certain that other new breeds of pedigreed cat will emerge. In time, pedigreed cats will become more common since they have the advantage of having defined breed characteristics that make them quite predictable in temperamental development. Another great advantage that these breeds have stems from the high level of appreciation the breeders hold for them. As a result, the kittens receive close attention including early social handling (ESH). By such treatment in the brief age period in which it is effective, the kittens acquire social capabilities that unattended kittens never fully obtain. Their social capability helps the pedigreed kittens to have very acceptable forms of temperament. Since problems of temperament lead to the rejection of a significant number of non-pedigreed cats, this could be seen as a desirable development from a welfare point of view.

3 Feline Neuroethology

When treated as a branch of applied ethology, the subject of neuroethology attempts to show connections between the main neural processes and elements of behaviour in domestic animals. The objective is to appreciate the links between these covert and overt factors in the dynamic functions of these animals. An improved understanding of the animal's mechanisms in its way of living can emerge to show the biological status of the subject. This can generate respect for an otherwise uncommunicative creature, such as the cat, which, in the past, was often dismissed as being merely reactive to stimulations. However, the cat has the essence of cognitive abilities in its style of living.

Neuroethology assists an understanding of the brain mechanisms that facilitate plasticity and adaption of behaviour to the environment. Some awareness of neurological controls is helpful in an appreciation of cat behaviour, as these relate to the diverse psychological functions of perception, memory, thinking and performance (Oliver and Lorenz, 1993; Bagley, 2005). The peripheral nervous system functions as an electrical communicator in which neurons conduct information towards and from the central nervous system. These neurons are classed as follows:

- Sensory or afferent neurons, which send messages into the central nervous system.
- Motor or efferent neurons, which conduct information out from the central nervous system.
- Interneurons or internuncial neurons, which are contained completely within the central nervous system and whose function it is to distribute and integrate information within this system. Implicit in this definition of interneurons is that they both receive and transmit information between adjacent neurons.

The nervous system's main parts are the central nervous system and the peripheral nerves (Dyce *et al.*, 1996). The central nervous system has its own major divisions, namely the brain and spinal cord. This system functions as a whole and all parts are interconnected through neuron linkage; only seven nerve cells are required to link any one part with another part of the entire nervous system.

The cerebrum, cerebellum and brain stem are the primary components of the brain (Sisson, 1995). In addition, there are special parts such as the cerebral cortex, the basal ganglia, the reticular formation and a range of specialized regions, including the hypothalamus. Specialized nuclei, such as the locus ceruleus, are also recognized in abundance. The locus ceruleus is the nerve centre for special attentiveness, as in prey watching. One notable system is the limbic system which is a collection of specialized parts. The brain stem is an important bridge in the processing and integration of behavioural output. Its notable structure is the reticular formation.

Reticular Formation

The reticular formation's functions include the production of general arousal in the cat's central nervous system. This structure embodies a mechanism by which states throughout the central nervous system are regulated. Some of these regulations are diurnal – one state is sleep and another is wakefulness. Between these two are many degrees of alertness and inattentiveness. All are expressions of some pattern of activity in the reticular formation, which is a place of convergence for information of widespread origin. This constitutes its role in ascending systems and also in a context of descent. Neurons in the rhombencephalic reticular formation can respond to inputs from secondary sensory cell groups in the spinal cord. They may also respond to a message from the cerebellum or the neocortex. A large variety and number of messages converge on the reticular formation, which has to process this excess of neural matter. It then dispatches impulses via reticulospinal fibres that terminate either on intermediate neurons or directly on motor neurons.

Cerebral Cortex

The cat's cerebral cortex is a unit with certain localized regions, where sensory impulses are received and subjected to redirection. These specialized areas of the cortex are primary sites for sensory reception; the nervous activity is subsequently spread over a wider area. The cortex has a multitude of cells and neural paths, each one communicating with many others. This extremely complex relay system permits tremendous variability in the way that nerve impulses may be channelled. The cerebral cortex possesses five main sensory areas into which projectory nerve fibres send signals. These sensory areas are as follows:

- the somasthetic or body sense area;
- the visual area;
- the auditory area;
- the olfactory area; and
- tactile areas.

The above cortical areas are important in the reception interpretation of nerve signals and execution of responses. The cortex is of fundamental importance in determining the cat's behavioural output.

Somasthetic area

This area is sited in the parietal lobe of the cortex and receives nerve impulses from many parts of the cat's body, particularly its complex integument.

Visual area

Nerve fibres are collected from the retinas of the eyes into the optic nerves and are distributed within the cortex to the extensive visual area at the occipital part of the cerebrum (Myers and Coulter, 2004). The recognition of set situations and the detection of releasers that activate innate patterns of behaviour take place in the visual area (Prince, 1977). The kitten's development of visual behaviour is dependent on experience (Norton, 1974; Blakemore and Van Sluyters, 1975; Villablanca and Olmstead, 1979).

Auditory area

The marshalling centre for hearing is located in the temporal lobe of the cerebral cortex. This area receives nerve impulses concerned with auditory sensations from the thalamus. The fibres of the hearing nerve end in the pons, from which region other fibres pass to the thalamus and then to the cortex for interpretation.

Olfactory area

The sensory area dealing with smell plays a much more important role in feline feeding behaviour than has generally been recognized. An olfactory region is located in the hippocampus which receives projection fibres from the centre in the olfactory bulb. This centre deals with olfactory reflexes (Sanchez-Andrade and Kendrick, 2009). The fibres of this system originate with nerve cells located in the mucous membrane of the nasal passage and terminate within the brain's olfactory bulbs.

Tactile areas

Certain areas of the cortex are designed to correlate dynamically with receptive areas on the integument. Superficial areas and central nervous areas change their importance according to the major activities of an animal. The appropriate area of the cortex is more receptive to stimulation, for example during predatory activities when the feline's concentrated attention is maximal (Fig. 3.1).

The Limbic System

The limbic system is a determinant and integrator of strategic and tactical functions, and also in displays of emotional behaviour. It acts as the central representative of the autonomic system, consisting of an interconnected group of brain structures within the cerebral borders. Included in this group are portions of the frontal lobe cortex, temporal lobe, thalamus and hypothalamus, together with certain midbrain parts that act as function generators (Heimer, 1978).

Neuron pathways connect all of these brain regions in segmental integration. The hypothalamus is the main output state of the limbic system and the components of this system have many connections, with each other and with other parts of the central nervous system. Information from the different afferent and efferent routes influences the limbic system and its peripheral arm, the autonomic system. The activity of the limbic system can result in a wide variety of autonomic responses and

Fig. 3.1. Housecat stalking prey.

body movements which comprise purposeful behaviour. Certain hypothalamic areas elicit behaviour which has a strong 'emotional' component. The medial hypothalamus, for example, exerts inhibition on the circuits producing the cat's fight-or-flight behaviour. Upon receipt of appropriate environmental stimuli, the temporal lobe inhibits the medial hypothalamus, allowing activity in the integrated limbic system to increase, with resulting emotional behaviour (Pryse-Phillips, 2009).

The structures involved in the control of emotional behaviour are predominantly located throughout the limbic system. The main controlling centres for consummative behaviour are located here, within the hypothalamus in particular. Much consummative behaviour relates to maintenance, the ongoing operation of which relates to behavioural homeostasis. Limbic matters therefore clearly relate to homeostasis, which is the cat's essential behaviour of self-maintenance.

The limbic system represents a device for providing the animal with means of coping with the environment. Parts of this system are concerned with primal activities related to predation and sex; others are related to emotions and feelings, and still others combine messages from the external world to the brain.

The limbic system is a fundamental regulator of survival responses on the basis of behavioural analysis. Each of its structures is highly specialized and tuned to specifics, either internally or externally, serving to regulate basic, primitive activities. The limbic system allows new directions in behaviour to occur; for example, when old patterns of responding fail to produce the anticipated rewards, limbic suppression then prevents the animal from continuing its old ways of responding and over-reacting in general. This system makes possible the suppression of established ways of responding in order to allow behavioural modifications based on information from the internal environment via neocortical tissue. It also gives the cat its own characteristic motivations, moods and emotions (Hogan, 2005).

The hippocampus and amygdala are two main components of the limbic system. The hippocampus is significantly concerned with emotion in behaviour. It is connected to the hypothalamus via a tract called the fornix to generate a physiological output in behaviour that expresses emotion. The fornix is another limbic component, located at the free edge of the cerebral mantle. Various parts of the forebrain convey neural activities to the hypothalamus through the limbic system, so that by one route or another, the hypothalamus is in receipt of impulses from optic, olfactory, acoustic and tactile sense organs. The limbic system also contains neural centres, such as the aforementioned amygdala, which control aggressive behaviour in its various forms. The frontal cortical areas play a role in coordinating the signals from the limbic system that are to be integrated with the activities of the 'cognitive brain', that is, the majority of the neocortical surface (Prosser, 1991).

The hypothalamus requires special identification. It can be regarded as an extended part of the limbic system, since it is its terminus. It also has an executive role as the hormone system's initiator and the autonomic system's controller.

The importance of the limbic system in well-being is major, although it operates subconsciously. Its function assists the cat in absorbing its circumstances and adjusting to the limit of its abilities.

Basal Ganglia

Three large subcortical nuclear groups are collectively called the basal ganglia and account for 5% of the brain mass. The basal ganglia nuclei, together with the cerebellum, participate in the initiation and control of physical movements. Many neurons of the basal ganglia change their activity during movement of a specific body part. These neurons are typically clustered close together, forming a somatotopic representation of the body. Neurons whose activity pertains to movements of the forelimbs are found ventral to those whose activity relates to hindleg movements. Characteristically, the changes in activity of these neurons occur prior to body movement, indicating that they play a role in predetermining the particular act. Their differentiation is of special significance in cats, whose forelegs and hindlegs have quite dissimilar functions. The hindlegs effect the feline's vital leaps, while the forelimbs are used in fighting and in catching prey. The claws of the forepaws are in much greater use than the hindpaw equivalents. The basis of certain motor patterns, such as purring, may be the dopaminergic projection from the substantia nigra to basal ganglia. Dopamine constitutes about one-half of the catecholamine in the brain, 80% of which is localized in the basal ganglia.

Hypothalamus

Behaviour that is largely dependent upon experience and learning in the animal is seldom, if ever, completely free of control by the primitive mechanisms established in the hypothalamus and in the subcortex (Forkman, 2002). At the level of the hypothalamus, patterns of nervous activities become integrated and regulated so as to establish the adaptive reactions of the animal. The neural links involving the subcortex and the hypothalamus, with the surrounding brain, remain the principal integrators of most behavioural patterns. The working units of the hypothalamus are neurons that are grouped into 'nuclei', which operate together with a high degree of coordination (Broad *et al.*, 2002). Information from various levels of the brain is received and processed by these nuclei before signals are subsequently reissued to more specialized parts of the body that are geared to function under the control of the hypothalamus.

The hypothalamus has a glandular appendage – the pituitary. It is also continuous in the forward direction of the limbic system. Much of the influence of hypothalamic activity is directed at the production of hormones in the subjacent pituitary gland. The pituitary is the principal endocrine gland in the body and its hormonal production is all-important in the maintenance of the bulk of the body's activities, including behaviour. Even the central hypothalamus is responsive to some of the endocrine activity for which it is initially responsible. Conversely, there is a considerable amount of hormonal control over the hypothalamus. The receipt of afferent stimulation gives the hypothalamus the role of maintaining and regulating the activity of the pituitary gland (Cunningham and Kline, 2007).

Autonomic System

Autonomic fibres passing without interruption from the hypothalamus to the autonomic motor neurons of the spinal cord's grey matter constitute a small minority of the outgoing hypothalamic fibres. The hypothalamus appears in large measure to project no further than the midbrain, where neurons of the reticular formation take over. The pathways descending to autonomic motor neurons are interrupted at numerous levels. At each interruption, further instructions enter the descending lines. The convergence of information on motor neurons is a characteristic of the autonomic nervous system.

Most sympathetic nerve endings release the neurotransmitter noradrenaline, which is chemically similar to the adrenaline produced by the medulla of the adrenal gland. The nerve fibres creating such chemistry are therefore termed 'adrenergic'. Thus, the adrenal medulla resembles a ganglion of the sympathetic system in its roles, releasing noradrenaline at the adrenergic nerve fibre terminations.

Parasympathetic nerve endings release acetylcholine at synapses and other nerve endings, and are therefore termed 'cholinergic' in their chemistry of neurotransmission. Following its synaptic release, acetylcholine duly spreads parasympathetic effects. The effects of the two divisions typically tend to be opposite and antagonistic – such antagonism is due to the chemistry of these divisions. Parasympathetic stimulation produces the lethargic effects of slower heart rate, lowered blood pressure, constriction of bronchi, and increased activity and vascularization of the gut. Sympathetic stimulation, on the contrary, prepares the subject for action. Notably there is an increased blood supply to the voluntary muscles. During moments of fighting or capturing and killing prey, wild cats assuredly have a widespread sympathetic discharge.

The effects of the parasympathetic system have a negative character, while those of the sympathetic system have a decidedly positive influence on well-being. Under normal circumstances, the two branches of the autonomic system tend to monitor each other, in order to create a neutral state of body and mind (Fig. 3.2).

This neutral mental state is enjoyed by the healthy cat during most of its daily existence. In contrast to a widespread sympathetic discharge is a parasympathetic discharge, which occurs in instances of severe trauma. Effects are comparable to those described above in cases of traffic accidents. While the injured cat may appear peaceful, it could actually be *in extremis* with shock. This state calls for emergency veterinary treatment involving fluid administration and other therapy as indicated by the case.

The autonomic system acts as a behavioural integrator – its full role is in modulating the intensities of behavioural responses, and in particular, the emotional component of behaviour. The autonomic system shapes the nature of reaction. To exemplify, cats play-fighting with an object illustrate true aggression. Thus, it plays a part in determining the nature of emotive states associated with behaviour. Through this system, conditioned reactions determine the nature of future responsiveness. Autonomic conditioning may last for years and requires little reinforcement. Any reinforcement to this system is powerful and may even occur in anticipation of a set of circumstances without them necessarily occurring. Autonomic conditioned reactions are fast and precede somatic components of defence, alerting, fight or flight. Autonomic innervation improves the acuity of olfaction, taste, hearing, touch and proprioception. In feline life, the autonomic system is involved in agonistic reactions, survival efforts and comfort-seeking and raking. The latter has a high priority in feline well-being. In raking, the cat extends its forelegs and extrudes its claws, hooking them on to a surface. It pulls on the hooked claws with alternating left and right forelegs in retracting actions that cause the claws to be plucked from their hooked places. This exercise is repeated several times in each session of raking (Fig. 3.3).

Spinal Cord and Reflexes

The brain and spinal cord are extensions of one another. Specific tracts through the spinal cord,

Fig. 3.2. Relaxed in a neutral state of mind.

Fig. 3.3. Raking; note extension of the upper spine.

to and from the brain, exemplify this scheme. The role of the spinal cord is sometimes autonomous, as in certain reflex mechanisms. Simple reflexes such as limb withdrawal or sacral reflex are in response to a local limited stimulus with cord organization. The croup reflex is an example of coordinated cord reflex action. In this action, the healthy cat raises its tail and hindquarters in response to its sacrum being stroked.

Specific motor reflexes involve not only excitation of particular muscle groups, but also inhibition of antagonistic muscles, as follows:

- Stretch reflexes involve excitation of extensor and inhibition of ipsilateral flexor muscles as occurs in pandiculation, which is of regular occurrence in cats.
- Flexor reflexes arise when painful stimulus to the limb causes excitation of ipsilateral flexor muscles and inhibition of ipsilateral extensors, causing limb withdrawal, especially the hindlimb when its paw is grasped.
- Withdrawal reflexes are associated with extension of the limb of the opposite or contralateral side of the body. Inhibition of the flexors on this side also occurs.
- Many neurons are common to all the above reflexes – ventral horn cells, for example, act as a final common path. These circuits are only part of more extensive circuits.
- In addition to allowing for many forms of integration, spinal interneurons allow a presynaptic input to be converted from excitation to inhibition or vice versa.

Inherited, complicated yet coordinated motor behaviour such as that which occurs in climbing can be observed in terms of all of the above mechanisms and reflex pathways (Jensen, 2002). The nervous system is responsible for innumerable varieties of co-coordinating, generating a continuum of behaviour which is directed at homeostasis in the cat's well-being.

Well-being Reflexes

Cat well-being can be inspected by testing for normal reflexes. Four such reflexes and their manner of execution are described below.

The sacral reflex

With the cat in its natural standing position at right angles to the examiner, the head is steadied by one hand while the other strokes the cat's back towards the tail. Passing over the lumbar area and the sacrum, the hand should apply extra pressure at the tail's base. With a positive response, the cat erects its tail firmly into an upright position. Some cats show an extra responsive feature by extending both hindlegs simultaneously. This reflex occurs in normal health. To confirm the response, the test can be promptly repeated several times.

The hindfoot reflex

With the cat in a resting position, the examiner holds one of its hindlegs below the tarsus and gently pulls that leg while slipping the grip downwards to the foot. In the normal, healthy cat the leg is quickly plucked from the examiner's hand as it slips over the foot. This exercise may be repeated one or two times before the cat rises and moves away.

The smiling reflex

With the cat gently restrained, the examiner places a cupped hand over the face so that the thumb and forefinger extend beyond the cat's vibrissae. The hand is drawn firmly backwards to sweep the whiskers along the side of the face. In a normal, healthy response, the cat will extend the corners of its mouth backwards, similar to a smile. As stated above, the test can be repeated once or twice before the animal begins to resist the examination.

The tonic immobility reflex

This test is a copy of the manner in which a mother cat carries her young kittens to a different location. In applying this procedure to a mature cat, the examiner seizes a large fold of skin on the back of the neck, close to the withers. The cat is then lifted with this hand, while the other hand is kept below the animal's hindquarters, without taking any weight. This placement is for security in the event of the cat slipping out of the grip. The cat is held up long enough to observe whether a state of tonic immobility is induced through the procedure. The animal is then placed on a solid surface, ending the test. The normal feline response is a clear indication of body firmness without struggling – flaccidity is not a positive response.

The tonic immobility reflex when lifted

When a mother cat transports her kitten, she clasps it by the loose skin over its neck and in front of its shoulder blades, using her incisor teeth. In immediate response to being lifted in this manner, the kitten develops a tonic immobility reflex, in which there is no struggle. The body is held stiff, the hindquarters are raised, and the tail and hindlegs curl up (Bessant, 2004b). As a result of this bodily reflex, the kitten becomes secured as a compact and balanced mass for carrying (Fig. 3.4).

This reflex is retained throughout the life of a healthy cat and is exhibited when the animal is lifted by the scruff of the neck for restraint or carriage. This method can also be useful for restraint in an emergency. If an adult cat resists brief lifting in this manner, it is likely that its state of well-being is compromised – perhaps through anomalous temperament or pain.

The value of these tests lies in the fact that the induced reflex indicates a normal state. Furthermore, the tests are simple to perform and are in no way aversive in their effects. Strictly indicators of general mobility, they are not to be performed as adequate clinical tests in the event that some pathological condition is under suspicion.

Fig. 3.4. Carried kitten in tonic immobility.

Pheromones

A pheromone is a chemical agent that provides communication between individual animals via the olfactory system. Like an external hormone, it is secreted and then released for detection in the central nervous system of another individual of the same species, in order to elicit a specific response.

The odorous substances of cats contain several pheromones. These pheromones are secreted not only through various sites on the skin, but also through urine, anal glands and other sources. Releaser pheromones can produce a prompt behavioural result, such as sexual stimulation. Primer pheromones produce a latent effect such as lodging a territorial claim. Evidently, a pheromone feedback mechanism stimulates sexual activity. Male cat urine contains an indicator sex pheromone. Pheromone detection depends on the cat's nasal ability to identify very fine odours. For this, the cat possesses the vomeronasal, or Jacobson's, organ. This is an olfactory receiver in the form of a pair of blind-end tubes within the nasal cavity, linked to the hard palate through a fine tubular system. The organ has a neural link going directly within its own mechanism of conduction to the centres of olfaction in the brain (Whitten, 1985; Hubbal *et al.*, 2009). This system depends on the cat drawing in a fine aerosol of air with pheromonal content, an action known as flehmen. In flehmen, the cat slightly extends its head, raises its lips and opens its mouth slightly. Pheromones detected via flehmen can signal the oestrous state of a female to a male and contribute significantly to sexual motivation in the latter.

The cat's two anal glands open on to the rim of the anus and are expressed with anal compression. They secrete an oily fluid with pheromonal property for the purpose of territorial scent-marking. This fluid, combined with excretory material, signals to other cats that the territory is occupied by an established resident. Deposits of faeces and sprays of urine are other principal means of territorial marking. Various pheromones contained within these deposits most likely indicate the status of the marker (Sunquist and Sunquist, 2002).

Another source of marker pheromones is the cat's skin. The skin areas involved with output of feline scent include the broad temporal area, the cheeks, the edges of the mouth, the flanks and the tail. The cat's scent is pheromonal material, which, like other pheromones, is not detectable by the human sense of smell. It is a continuous occupation of a cat to mark items in its environment with these pheromones by pressing its body areas of source against them. This action marks these items as possessions of that particular cat. By nature, cats are very possessive, and display this tendency by rubbing the side of their mouth, head and body against their owners and other people. Cats live in their own scented milieu of which humans have no experience. This can create deficiency in human comprehension concerning the cat's values and messages contained in its scented communications.

Hormones

Hormonal effects on feline behaviour are obvious in various situations, including the sexual activities of a mature male cat (under the influence of testosterone), the periodic outbursts of intense activity (from surges of adrenaline), the oestrus display in the female (from oestrogen), arousal (vasopressin) and within general activity (thyroxine). These hormones function as circulating messengers carrying biochemical signals to systems of the body, including neural systems that organize behavioural output by the motivation of activities that are hard-wired within the central nervous system. Although most hormones originate from the ductless endocrine glands such as the pituitary, the thyroid and the adrenals, certain releasing hormones proceed from the hypothalamus, which is controlled by some of its antecedents in the limbic system. This shows that the brain has ultimate determination in the hormone chemistry of the body (Feldman and Nelson, 2004). In consequence, a state of well-being, including mental health, is basic to the proper orchestration of the endocrine system. The latter has its own part to play in health support.

By way of the corticotropin-releasing hormone, there is output of adrenocorticotropic hormone (ACTH), which is important in the management of stress. It causes the cortex of the adrenal glands to produce various corticosteroids, such as cortisol. ACTH is the essential hormone for adapting the animal to unusual circumstances, increasing its output in stressful matters of housing or management. Steroids are produced in the adrenal cortex as hormonal substances. The adrenal cortical hormones can generate glycogenesis by glucocorticoids and male activity by androgens. All of the adrenal cortical hormones are termed 'corticoids', including cortisol, which activates carbohydrate metabolism so as to increase blood sugar for sudden energy to deal with stressful events, such as an unexpected fight. The adrenal gland's central medulla is under the control of the limbic system's hypothalamus, and with stimulation from this origin it produces adrenaline – another hormone that can prepare the cat for emergency with impulsive physical activity. Under motivation, this hormone can quickly create increased vascularization of musculature for exertion (Hogan, 2005).

It is evident that the central nervous system and the adrenal system have a working relationship – they influence each other by secretions. Hormone secretion is subject to the influence of many forms of stimulation and there is an elaborate organization of interactions between the external stimuli received and the internal physiological state. Behaviour, environment and internal state can alter to cause a change in motivation. This can create complex situations of chemical activity in neuropeptides, hormones and pheromones, all of which are active in chemical communication within the body. Such communication advises and instructs the animal's parts on the physiological status and requirements of the whole (Schulkin, 1999). This active chemistry provides a powerful basis for the cat's most essential activities (Fig. 3.5).

Fig. 3.5. Arousal in hunting.

Autonomic Network

The two divisions of the autonomic nervous system are largely antagonistic in their differing functions. The parasympathetic division is chiefly concerned with the basic control of the internal organs. The sympathetic division can exercise a dominant effect, particularly under stressful conditions, as when the sympathetic nerves accelerate the heart. Other functions require brief mention, as follows:

- The parasympathetic nerves have acetylcholine as a neurotransmitter to relax blood vessels, stimulate gastric juices and tone musculature.
- Sympathetic connections go to the medulla of the adrenal gland (which produces adrenaline), sweat glands, skin and the capillaries of voluntary muscles.
- Noradrenaline is a sympathetic transmitter and has its derivation in transmitter dopamine. Neurotransmission utilizes significant amounts of calcium and potassium – perhaps more so under stress.
- Sympathetic nerves react to stressors. In stress, noradrenaline serves to stimulate the secretion of adrenaline into the bloodstream. Adrenaline is thus an emergency proceeding to all parts of the body and stimulating major bodily action.
- Disturbed physiology in the hormonal and neural autonomic network influences certain hormones and transmitter substances such as adrenaline and dopamine respectively. Severe or chronic imbalance in this elaborate network underpins stressfulness in the cat's behaviour.
- The autonomic network is of such fundamental importance in maintaining well-being that it could be regarded as its foundation.

Neurotransmission

The cat, like other higher animals, has a neural network in continuous operation to enervate its living mechanisms, including behaviour. This network is comprised of neurons that communicate by the release of neurotransmitter chemical substances. The central nervous system organizes this process while dealing with sensory discrimination, stimulation, innate motivation and behavioural output. The executive arm of each neuron is the axon, which communicates its chemical substance to the receiving extensions, the dendrites, of the next neuron in the network. This linkage occurs at the synapse, the junction between two neurons, where neurotransmission occurs to create electrical patterns. These mobilize the cat at one end and provide consciousness to the animal at the other end of the neural network. Behaviour thus uses an electrical principle.

Neurons are of two main types. The first type is the macroneuron, characterized by long axons that make up the primary nerve tracts. These neurons are very stable in their physiological work. The other type is the microneuron, which is present throughout the nervous system but most prominent in areas of the brain (such as the cerebral cortex) that deal with alterations in living circumstances. Microneurons develop after birth and are responsive to a kitten's experiences (Villablanca and Olmstead, 1979). Small groups of neurons spread their axons with amine fibres throughout the central nervous system and act as a system of higher command. The transmitters that function in this command system are the aforementioned noradrenaline, dopamine and serotonin.

Depending to an extent on experience, neurons make permanent linkages where appropriate, and lose other connections with target areas that are not drawn into functional transmission. This manner of practical/empirical neuron coupling uses information from the particular environment of each cat to shape its ability to function in that environment. To illustrate, kittens that do not experience social handling will not develop the same quality of human affiliation in later life as those that are well-handled. Thus, environmental conditions introduce variability into the development of the neonatal nervous system. Additionally, kittens have a functional ability prior to sensory capability, so that essential actions, such as nursing, may be performed without sensory interferences. The newborn kitten's deglutition is programmed to function by a combined action of sucking and swallowing. If the kitten is orphaned and is fed by an eyedropper or syringe, it quickly becomes able to ingest without the sucking component. Another example of neuroplasticity is recognized in congenitally deaf cats (such as blue-eyed whites) that have superior eyesight as a result of their olfactory cortex aiding their visual cortex in sensing the location of movement.

The transmission of signals at the synaptic junctions involves four steps, as follows.

1. Synthesis of the transmitter substance in the neuron cell.
2. Transport of the chemical along the axon to the synapse.
3. Action of transmitter across the synapse to the receiving membrane of the next neuron.
4. Removal of transmitter from the synaptic cleft by absorption, degradation or re-uptake.

Transmission may be fast or slow, depending on the circumstantial nature. To exemplify, the cat will require fast transmission for the production of urgent action, and slower transmission for rest and sleep. Slow transmitters are the catecholamines, noradrenaline and dopamine, while acetylcholine is for quick transmission. Most neurons use only one transmitter substance. The four types of transmitters are as follows.

1. *Acetylcholine*. This is the transmitter heavily involved in the function of the motor system. It is also the transmitter for the parasympathetic system.
2. *The biogenic amines*, including dopamine, noradrenaline, serotonin and histamine. Dopamine and noradrenaline are important transmitters, utilizing common pathways. Noradrenaline cell bodies are prominent in the locus ceruleus in the brain stem, the nerve centre for arousal. Noradrenaline is also the transmitter for the sympathetic nervous system.
3. *Amino acids*, namely γ-aminobutyric acid (GABA), glycine and glutamate. Amino acids exist both as cellular constituents and neurotransmitters. Glycine and glutamate are two of the 20 common amino acids that are incorporated into the proteins of all cells. GABA is present in the basal ganglia, the cerebellum and the spinal cord.
4. *Neuropeptides*. Numerous peptides exist among neurons and are very active in the causation of inhibition or excitation. Some of these peptides function as hormones by neurosecretion, such as oxytocin, vasopressin, luteinizing hormone and thyrotrophin-releasing hormone. The neuroactive peptides are concentrated in parts of the brain concerned with the perception of pain, pleasure and emotion. In addition, two classes of peptides, namely the endorphins and the encephalins, have opiate properties that modulate pain. The most active endorphin is synthesized in the hypothalamus and the pituitary, while the encephalins are synthesized in neurons and the adrenals. Substance P involved in pain sensation is a peptide in the

Fig. 3.6. Cat exhibiting normal feline curiosity.

basal ganglia, the hypothalamus and the cerebral cortex.

Motivational mechanisms within the brain arise from peptide activity (Hogan, 2005). Curiosity, for example, is one of the major motivating factors in the cat (Fig. 3.6).

The biogenic amine system is involved in negotiating the cat's homeostasis. Noradrenaline and dopamine control some of the motivational hormones. The dopamine pathway is involved in avoiding noxious stimulation and carries incentive messages. Incentive rewards, such as the smell of food, utilize dopamine pathways while final rewards, such as completion of feeding, act through noradrenaline pathways. Motivation in the cat is related to environmental factors and endogenous factors, such as hormones. The principal messengers in motivation are neuropeptides and hormones; both of which have phasic production, giving rise to phasic motivation such as hunting, feeding, avoidance, resting, body care, exploration and affiliation (Insel, 1992). Collectively, these activities, stemming from neurotransmission, establish homeostasis in the cat, thus maintaining a steady state of well-being.

4 Play and the Steps Through Life

Play

The first organized activity of feline life is play. Kittens play with much energy and their playfulness characterizes this animal. Play continues into adult life (Hall, 1998), with diminishing commitment as the cat gets older such that the degree of playfulness shown by any cat can be an indication of its stage in life. Nevertheless, play is a cat's special feature that requires examination as a behavioural phenomenon. Play is very much a part of development from the very young kitten onwards to adulthood. Cat play can be defined as an activity that has feline behavioural components and emotive quality, but is unrelated to any immediate need of self-maintenance and targets a known conspecific, or a movable object, for feedback that facilitates continuity of play (Fig. 4.1a and b).

Nature of play

The kitten has to learn its physical capabilities and a variety of tactics to live as an adult. Many of the learning activities take the form of play. A wide range of unnecessary actions, such as fighting (Pellis and Pellis, 1998), that are performed by kittens and young cats providing them with physical exercise can come under the heading of play. Such actions include the following neuromuscular rituals:

- making body signals to initiate or prolong play with a sibling;
- fighting with a conspecific, without causing injury;
- chasing, fleeing or hiding in connection with an affiliated conspecific;
- advancing and retreating, or posturing, in the face of a conspecific mate;
- performing acrobatic exercises such as leaping upward or forward and body twisting;
- moving and manipulating objects in its environment; and
- repetitions and extensions of any of the above forms of behaviour.

Many forms of kitten play are social activities involving intimate contact with a littermate or the mother. Play among several kittens in the litter aids in the cohesion of the litter (Mendoza and Ramirez, 1987). Social play in paired situations usually ends with mutual satisfaction in physical output. Playfulness is more evident in larger litters (Mendl, 1988). At other times the kitten will perform solitary play, which provides exercise and apparent satisfaction in expended energy. This aids the kitten's development of motor activity (Levine *et al.*, 1980; Martin and Bateson, 1985). It also trains the animal for the unexpected (Spinka *et al.*, 2001). The accompanying emotions in playful activities have a realistic nature. Anger, fear, fright and affiliation can be shown.

Functions of play

The experiences acquired in play facilitate learning, social bonding and inhibitions. Physical development is an obvious gain in play. The increased circulation to the involved musculature and the increased vascularization in general advances health in the growing cat. Since the patterns of play usually involve exaggerated and repeated movement, a boost to cardiac function results. Partners in play often change roles in the course of the game that is being played. Insofar as play behaviour assists development in the young cat, its encouragement can be regarded as a welfare need.

In the litter, social play begins in the fourth week and reaches a peak of activity in the seventh week when object play enters the interest of the kittens. Play in the litter has an aggressive nature by the fourth month and true fighting starts to occur then.

(a)

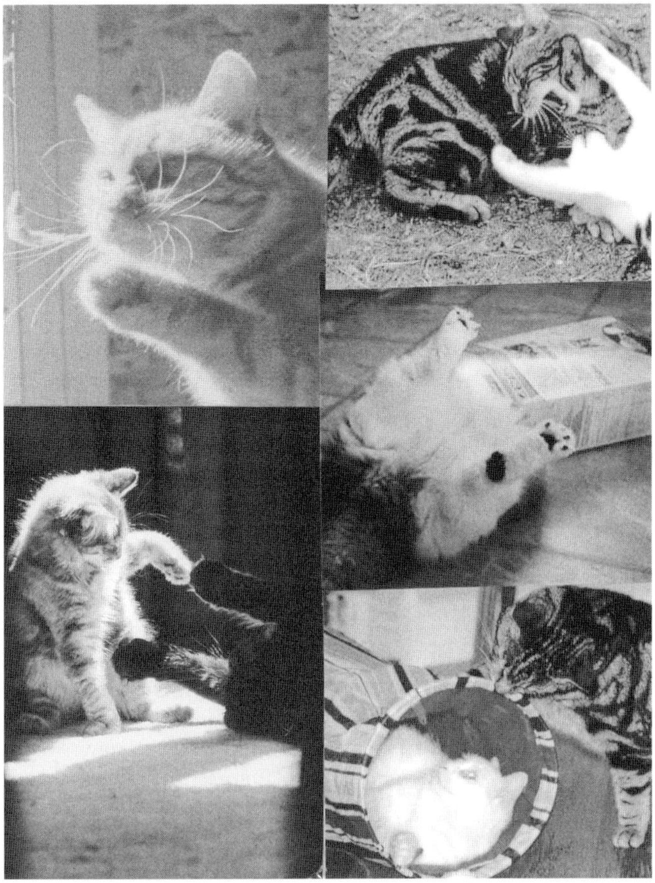

(b)

Fig. 4.1. (a) Differing forms of play; (b) a friendly fight – note expression on face of the 'belly-up' defendant.

This corresponds with the behaviour of dispersal seen in the wild cats, when the litter breaks up and the subadults leave the natal territory to establish home ranges of their own.

In cats, the motivation and physical performances in play are inherently the same as in real acts of predation. The reactions to reality and to its play version have apparent neurophysiological unity. The main difference between them is that in play, there is no true consummation, while in real predation there is finality in the performance. In the infant cat play is part of its epigenesis, while play in the adult is displacement activity arising from an absence of opportunity to discharge an inherent motivation for predation.

Play becomes a major activity as the young kitten develops and at first it is often directed towards a sibling, the mother or an object. As the play becomes purposeful, the object must be movable and mobile to provide a result to the action of play; the kitten's play requires a target and a result, or response. Play with siblings gives maximum response in mutual action. The kitten is commonly regarded as a most playful subject and this characteristic continues from this stage throughout adult life, typifying one of the cat's desired features as a companion animal. Play is pursued with real emotions such as aggression, fright, anger and affiliation. Such emotions are interchangeable with the partner and can also be directed at the chosen object. The nature of play in the cat will be discussed further, but three of the kitten's play actions appear in this phase of development and are as follows.

1. *Manipulation.* Actions of a manipulative nature are prominent in cat play. These include ambidextrous flicking movements with alternating forefeet, catching the play item with the mouth, tossing the play item sideways out of the mouth, trapping it with both paws, poking and batting at a target with a forepaw.
2. *Posturing.* During active play, the kitten may halt briefly to adopt a special posture that leads to a given action. This includes an upright stance on the hindfeet and tarsal joints for stability. The forelegs are extended forward in this posture, which usually initiates a playful surge forward.
3. *Kinetics.* Episodes of play typically include sudden, vigorous acts of running, leaping and strutting. These kinetic features include running stopping and running; various speeds of chasing and fleeing; pouncing; strutting sideways with an arched back; upward and forward jumps; stealthy creeping.

Playing requires energy and it is therefore only seen in generous volume in kittens that are well nourished. It is a manifestation of well-being. The kitten's caregiver can encourage playful activities with the use of suitable play toys. Such exercises aid the development of the kitten.

Phases of Feline Development

A state of well-being can exist in any age of the cat, despite physical and behavioural changes that occur through its progression from one age group to the next. Modifications in the animal's care become necessary as it gets older.

In an examination of the phases in domestic feline development it is necessary for discussion to define such periods, as these are perceived to be in accord with the broad feline picture in this text. The margins of these phases are somewhat arbitrary; they are not fixed and they may alter or overlap in the reality of biological variation, but they help to show the main steps in the progress of development as they typically occur in the young feline subject. Four of these states, or phases, are recognized here, as follows.

Neonatal phase

The neonatal state is applied to the animal from birth to the multi-sensory period when the kitten has acquired some sense of vision, hearing, directional awareness, the ability to stand and some coordinated locomotor ability (approximately 14 to 20 days of life).

As it is born, the domestic kitten is blind, deaf and unable to walk, but it is not without certain sensory faculties or specified motivation that begin its behavioural epigenesis; physical developments can then follow. The initial motivation is to struggle with its forefeet to the mammary gland. The mother exposes this after the last kitten has been born and licked clean of fetal membranes, all of the latter she consumes. She uses her carnassial (i.e. the meat-shearing) last premolar tooth in the upper jaw (that contacts the lower jaw's molar), severing the umbilical cords. At this time, the neonate has an average weight of 100 g that will be doubled by the end of the first week if the mother is well fed and lactating properly.

A neurological factor in the neonate creates attraction to the mammary area through a mammotaxis, which provides direction, snout first. The clues are the warmth and odour from the relatively hairless mammary region. On contact with the mother's ventral anatomy, the kitten nuzzles around that skin to locate precisely the mammary gland and one of its nipples, which is now protruding as a result of the localized stimulation to the mother. The neonate cat now rubs its snout up and down into that focal point repeatedly.

The mother is already primed for this initiating event, through her pituitary hormones prolactin and oxytocin; her colostral milk exudes from the teat with some pressure with the 'let-down' effect of the latter hormone. The newborn kitten now sucks and swallows the liquid eagerly, consuming it until its stomach is replete. Its infantile neurons for sucking will be lost after weaning, since adults do not suck.

Littermates quickly learn the location of their own teat, the one they initially used. Each continues to use the same one from the second day onwards, although they may use more than one as they mature and develop more appetite. Feeding times become organized by the mother calling with the 'chirp' sound and purring when she wishes to feed her young. The young learn to massage the teat area to stimulate the mother's feeding cooperation, including the 'let-down' of milk with the assistance of the hormone oxytocin. The kitten can purr by the third day and usually purrs while sucking the mother. By the fourth day the kitten can crawl a distance about the length of its mother. The neonate's hearing is established by about 2 weeks of age.

The simple programme of neonatal behaviour for approximately the first 14 to 18 days consists of feeding for a few minutes, then sleeping for about 20 min, then repeating the feeding procedure. The total time spent feeding at this stage may extend to 8 h (Bessant, 2004a). While sleeping, the neonate remains in its usual location at right angles to the mammary gland, sometimes holding on to it with its mouth when the mother rises. Littermates lie against each other during sleep. The mother usually initiates the feeding sessions during the phase of neonatal life. The reverse is the case during the following sensory period. The kitten can usually crawl beyond the birth site by the start of the second week.

A principal objective in the life of the newborn cat is to maintain its own heat. Having left the heat of the uterine interior, it is now vulnerable to heat loss. The mother provides warmth by her close presence. A good den, or a maternally well-chosen birth site, provides some insulation. Should heat loss occur at this stage, hypothermia is likely to become established and this leads to loss of vigour and arrest of feeding, with death resulting. Mortality rates can be about 20% in feline litters as a result, principally, of hypothermia and inanition. Both in the wild species and the domestic one, there is an additional, notable cause of death among neonates, namely cannibalism by an intruding male cat.

The eyes open at variable times within this phase and usually before the end of the second week. The neonate does not have good vision at this time, but when the aqueous humour clears in a few days, it can see more clearly. Its visual accommodation improves with the further exercise of the elliptical, muscular, ciliary bodies controlling the pupil.

Sensitive phase

The sensitive phase is taken to be the period from the start of multi-sensory development to the early post-weaned period (approximately 21 to 60 days, or 3 weeks to 2 months).

The sensitive (or sensory) phase is begun when the kitten can see properly. Vision continues to improve in this period with the aid of functional optical work, such as efforts at accommodation and experiences of differing light levels plus, and most importantly, focusing on moving images. It can now begin to relate to its immediate surroundings. Vision is progressive and is the essential factor in this stage. The sense of hearing becomes more functional after 2 weeks of age and, like eyesight, its development is progressive for a further 2 weeks or so. Thermogenesis is developing at the start of this period and by 1 month of age it is fully functional, giving the kitten the opportunity to move away from the warmth of the mother and den when it is able to walk at about the fourth week. It can recognize its mother by sight at this age. The adult forms of locomotion, including running, are developed by the seventh week. Also at this time, the kitten shows its ability to choose food. It rubs its forehead, cheeks, flank and tail against the mother from time to time, engaging her attention. When the mother brings prey to the den, the kitten shows an ability to manipulate it by 2 months of age.

An element of cognition progresses through this phase, allowing the kitten to make some comprehension of what it experiences. This is the time of a consciously intimate connection between kitten and the mother, creating the necessary situation for learning in the kitten by coaching from the mother. Cognition is implicit in the kitten when its behaviour actively and suitably relates to its ambient circumstances, such as knowledge of its placement and management of its self-defence. Repetition of any such behaviour indicates learning. Producing a copy of the mother's behaviour, such as the use of litter box or choice of food, is stronger evidence of learned behaviour. Coverage of faeces, however, is innate behaviour from the feline genome, but the maternal example strengthens the habit. Both of these forms of behaviour, namely food choice and faecal burying, become evident in this critical phase of development.

The provision of benign experience in this phase, such as human contact and, in particular, increasing durations of handling, has an enriching effect on the kitten's savoir faire, namely its ability to act suitably in situations it has previously experienced and survived innocuously. However, the pre-existing, inherent disposition of the individual kitten can be a modifying factor; it can alter the nature of whatever tendency is formed from such conditioning. The behavioural reserve following such prescribed exposure cannot always be predetermined. Nevertheless, opportunities to gain desirable conditioning should be afforded the kitten in an effort to improve this quasi-domesticated animal's well-being in its future role as a companion animal.

The particular conditioning is termed here as ESH, 'early social handling'. It needs to be carried out by frequent, careful and sensitive handling of the kitten. The animal should be frequently lifted and held close to the handler for fairly brief periods so that the animal becomes accustomed to this experience but not distressed by it. The kitten becomes customized to human association fairly quickly and starts to show a relaxed and friendly attitude to its handler from the experiences of humane handling (Fig. 4.2). This emotional state is carried into its adult phase, making it a very manageable companion animal.

In this context, the hypothesis is maintained that the kitten has an inherent disposition of a given type from its parentage and that experience can blend with this to establish the nature of the animal's unique temperament. It is further considered

Fig. 4.2. Conditioned kitten showing adjustment.

that the formation of temperament begins in this sensory period in addition to the formation of some goal-directed behaviour, such as some forms of self-maintenance. In common use to refer to temperament are the terms 'personality' and 'individuality' and these can be taken as interchangeable and synonymous with temperament's manifest 'style'. Some of this may be apparent in the kitten's manner of play as this begins in the later part of the sensitive phase. The kitten's weaning by the mother is contained in this stage. Weaning will normally have been completed at some point close to 6 weeks of age in the kitten and sometimes before this.

Percipient phase

The percipient phase is taken to be the time when the kitten focuses its perception on the mother's behaviour, learning specific abilities from her (approximately 2 weeks to 6 months).

With the foundation of the animal's ethological constitution duly laid, the continuing epigenesis of the individual goes further into a constructive stage of basic cognition, such as recognizing mother and siblings. In this stage, the kitten is learning continuously and imperatively how to live on its instincts, its knowledge and physique. For this purpose, it perceives with all its senses the features of daily events that are significant to its interests. In large measure, the kitten focuses on maternal examples of action, such as those that are food-related, and it copies from these.

In addition to copying, learning results from what is perceived as effective behaviour, for example during general play in which physical capabilities are learned. The great flexibility of all the vertebral joints, and the vertebral column itself, is an important factor to be learned and used. Elements of predatory behaviour emerge in play at about 1 month of age.

Subadult phase

The subadult phase is taken to be the period from the start of puberty to the time of solo existence, in accord with wild cat behaviour (approximately 6 months to 1 year).

This phase represents the young cat's adolescence. If the kitten has not been neutered, it spans the period of puberty. For all young cats, it is the phase in which they mature in behaviour and physique to the point of being psychologically independent and able to live alone. In the wild it is the period in which the young cat learns how to live by hunting, killing and appreciating its ecological state. This requires an extended apprenticeship under maternal tutelage. In many wild feline species this continues into the adolescent cat's second year, before the behavioural feature of dispersal occurs.

Prior to dispersal, the wild cat mother with a litter remains in anoestrus and any female kittens also fail to show oestrus in this phase. Evidently cats in the wild have an asexual predispersal period. This feature of asexual adolescence may be due to some inhibitory factor on ovarian function. It is one of the main differences between the wild and the domestic species. Sometimes, especially if her kittens have been removed from her, the domestic queen may show oestrus as early as 4 to 6 weeks after the birth. This predisposition to early breeding postpartum permits many breeding queens to have three litters per year.

The domestic kitten has a very different adolescence compared with its wild relatives as a result of its early maternal separation. For example, snow leopard cubs are still with their mother as 1-year-olds and do not attain their sexual maturity until they are 2 to 3 years of age. In the Eurasian lynx species, kittens are still seen with their mother as 2-year-olds, hunting and travelling with her. With the serval cat, the mother actively chases her kittens away from her when they are about 1 year old. Young ocelots start hunting with their mother when they are 3 months old and remain in her company until they are 2-year-olds. All the large cats keep their young with them until they are well grown.

The lack of adolescent tutelage from the mother is one of the principal effects that domestication appears to have had on *Felis catus*. The loss, however, facilitates customizing the kitten for the specifics of its novel domestic circumstances. This may have been so from the beginning of their domestication. It has always been the custom to take kittens away from their mother as soon as weaning has occurred and then to raise them alone in human custody. With this as standard management, it is possible that maternal tutelage became lost in the evolution of the domestic species, or it may only have become suppressed by customary management. Since it facilitates early re-breeding, however, the resultant excesses in kitten production that occur have created one of animal welfare's major problems.

Another consequence of the absence of ecological tutelage in adolescence in the domestic cat is that it is functionally incompetent in self-maintenance by predation. Although it has the innate instinct for predation, without seeing it performed by an experienced teacher, it has not learned the efficient methods of practising it as a way of living. The opportunistic killing of a rodent or bird in its limited home range is not an indication of an ability to kill wildlife on an adequate scale to survive if suddenly placed in unfamiliar territory. The domestic cat is born as a potential predator only, but cannot be a full-time operator of the art without maternal training. It is therefore an act of cruelty to abandon a household cat in some rural location to live by hunting for its own food. In reality, such abandoned cats soon die of starvation. The feral cat is not the same as a domestic cat, chiefly through lacking ESH and being permanently homeless; as a different subject it will be discussed later.

Seven Categories of Development

A parallel can be drawn between the cat's age status and The Seven Ages of Man, described by Shakespeare. The ages of the cat can similarly be put into seven categories as follows: infancy, juvenility, youth, maturity, middle age, old age and extreme age. These phases of a cat's life merge together without precise margins.

Infancy

When a kitten has survived its initiation into suckling, it can be regarded as a feline infant a few days after birth. At first, the kitten is involved entirely with self-maintenance, suckling on cue, crawling and huddling close to its siblings and its mother, taking its fill of milk, sleeping between feeds and existing in harmony with the composite litter. As the kittens suckle, they knead their mother's abdomen with their infant claws that become retractable when they pass their first week.

During their second week of life they open their eyes and have control of their claws, extruding them to knead and retracting them to sleep close to their fellow kittens in the litter. The young felines now begin relating to each other, participating in playful acts. As play improves, the entire litter partakes in little contests. Before long they can walk about freely, but will remain in the nest area.

Various gymnastic actions are performed by the infant kittens as they grow, notably leaping upwards, pouncing on each other, jumping forwards and creeping intently. These actions mimic the tactics employed by adult cats in hunting. As a result of their scrapping sessions it is usual for one kitten to emerge as the dominant one by the time they are a few weeks old. At this time it is important for the kittens to experience ESH, as a prelude to becoming good pets in later life.

When the kittens approach 6, 7 or 8 weeks of age, weaning becomes imminent. The mother no longer tidies up their excretions and the exchanges within their sessions of play become quite aggressive. The less dominant young even learn to fight. When the finality of weaning arrives, the mother rises when any kitten attempts to suckle. She steps over her litter and moves away from them.

Solid food should be provided prior to weaning age to create a gradual transition from one diet to the other. Once the young have ceased suckling and exhibit detachment from their mother and security in their independence, they have outgrown infancy and have entered their juvenile state, with males outgrowing females. At the onset of this stage, the young felines are eating sufficient amounts of their own food.

This is the age when a social capacity must be stimulated by ESH. This time for socialization is short and when the phase is terminated by other, superseding, epigenetic events, the social facilitation is ended, leaving the animal without that feature in its personality. This is a great loss in the animal's behavioural equipment for the remainder of its life. Modifications to any subsequent human relationship can be created by habituation, but the essential affinity for human company that ESH creates is never fully achieved.

Juvenility

Between the post-weaning age and puberty, the kitten is regarded as juvenile. Throughout this phase, the chief features are physical and social developments. From a well-being point of view this is a danger zone, as the kitten has now lost the immunities it obtained from its mother's first milk. It is inevitably removed from the seclusion of the birth site and put into an entirely different environment. Although its own immune system has started to function, the juvenile kitten is faced with health challenges due to its susceptibility to serious infections resident in the broad population of cats.

Even if the juvenile animal is being reared carefully as a house-bound cat, it will continue to remain exposed to feline viruses that may, or may not, respect closed doors. Outside these doors, the growing cat is exposed to all the hazards of infection, infestation and injury. Vaccination is vital to all cats in this age category and anti-parasite treatment should be scheduled routinely. A commonly recommended age for neutering is 4 months. Neutering after puberty has started is less satisfactory.

The ESH should now be supported by affectionate contacts on a fairly frequent basis. Even if these contacts are brief, they serve a psychological purpose by compensating the socially sensitive kitten for its loss of contact with its mother and siblings. Teasing the kitten into play with lures or toys is similarly helpful in drawing out the juvenile activities that are natural to the animal in its physical and behavioural developments.

Juvenility is the main period for learning. If the kitten has not yet been introduced to a litter box, it must now be provided with one containing good quality, fine litter to a depth of at least 5 cm. Learning to use a litter pan comes easily to a juvenile cat, as this behaviour is well implanted in its nature. A kitten that has not adopted a litter box immediately can be easily trained to use it, by repeatedly placing the animal on top of the clean litter whenever it begins to adopt its excretory posture.

The use of a scratching post can be taught by placing the kitten at the post during the initial onset of scratching. This should be repeated as often as is necessary to complete the training. An outlet for the cat's vigorous, innate clawing exercise (VICE) is necessary for the animal's well-being. Kittens can be trained to desist from unwanted activity, by promptly spraying them with water from a distance when they are performing the act that is to be prevented. A water pistol or large syringe can be used for the purpose of directing a fine jet of water. If the jet is sprayed from a distance, the cat is unlikely to associate the discipline with its trainer or caregiver.

By this simple water method, cats can be trained to stay away from houseplants. The water jet method can also be used to prevent the kitten from scratching furniture and curtains. The juvenile kitten must be trained before such bad habits become established. The training should be started as soon as the juvenile cat arrives in its new and permanent home. Discipline can be carried out through vocal discouragement or water spraying. Lessons are not well learned, however, when the cat has a hyperactive episode, which is more common to occur in extended kittenhood. Any tendency of the kitten to bite its handler can be quickly discouraged by the handler blowing sharply on the young cat's face.

Youth

The cat enters its youthful phase at the start of puberty. The first sign of heat in the female is the clearest indication of this period. While 7 full months is a common age in cats for the process of puberty to begin, there is a wide range of ages for this development. Female kittens have been known to come into heat as early as 7 months of age, while others as late as 14 months. Arbitrarily, 7 to 8 months is given here as a rough estimate of the average time when the gonads begin to function in both sexes.

During the youthful phase of a cat's life the animal is still in danger with regard to health and habit. All the necessary vaccinations should have been completed by this time. The other great concern is over sex. The youthful female cat is too young for pregnancy, and the male youth should not be allowed to develop strong masculine habits and personality that would be unsuitable for its domestic role. Neutering for both sexes should therefore precede puberty at its average time.

Kittenhood is only a shadow of the cat's new chronological status. Curiosity now gains momentum. Adventurous activities are undertaken more vigorously, both indoors and outdoors. These activities include such extravagant behaviour as climbing trees, shadow boxing on hanging objects, racing through the home or across streets, playing with maximal vigour that often resembles manic activity, fighting seriously with associates, pouncing on moving articles, exploring dark recesses and taking a keen interest in the great outdoors.

In addition to self-maintenance and energetic pursuits, the youthful cat engages in more gentle occupations such as grooming a mate or caregiver, seeking to hide in receptacles, occupying look-outs and taking up the common feline routines of resting and sleeping, at length.

The youthful cat has a good appetite and may consume edibles made available that are unsuitable in type or quantity, which in turn can contribute to an unhealthy state. It is the caregiver's responsibility to manage the cat by proper feeding, so as to protect the feline from itself. At this time, its growth is being consolidated and proper nourishment is needed to accomplish its full physical development. At about 1 year of age, the young cat has now stepped into maturity.

Maturity

The mature stage of many cats is a long period, extending from approximately 1 year of age to over 6 years of age. This is the phase when the adult cat is at its peak in terms of health and capability to cope with challenges in its life. The generalization that the mature cat has good, stable health is based on the assumption that the animal has received and continues to receive all preventive health measures, that it is a house-bound pet and that it has been neutered. Its secure existence is like a permanent quarantine.

Physical maturity provides the adult cat's body with strength, through frame, muscles, claws and teeth. All the internal organs are likely to be entirely functional in their physiological work. A healthy coat, well-groomed by cat and caregiver alike, will reflect the animal's general soundness. The mature cat's behaviour is like a symphony of the activities previously described individually as the dynamic components of self-maintenance, while its life may continue on a plateau of well-being.

The satisfactory condition of the healthy mature cat cannot be taken for granted. Continuous monitoring of the animal's appearance and its various features is a constant duty of the caregiver. The text of this book provides all the knowledge needed for such monitoring. The usual daily attentions to the housecat in its general care and the common affectionate contacts, however, make a very adequate system of monitoring the animal's well-being.

Occasional, physical interferences of the mature cat are not to be overlooked in its care. Tufts of hair may have to be removed, dirty ears may call for cleaning, claws may require clipping, teeth might need to be de-scaled, vaccinations may need boosting and parasite control could become necessary. Clinical interventions are more likely needed by cats on day-release from home, but clinical attention is never completely removed as an emergency need for any cat in any age group.

Outdoor, unneutered cats often have a shortened life expectancy and many of them do not live far into this phase of existence. Sundry types of accidents often befall them, including road-kills and loss from home. Suffering goes with these misadventures. City cat owners who wish to allow their cats the total freedom to roam outdoors may have good intentions, but should do so where road traffic is limited.

Without the option to return to 'the old country', the notion that complete, environmental freedom is favourable for Western, urban cats is seriously flawed. By and large, many such cats do not enjoy their prime of life to its full extent by living for 6 full years – half of the common life expectancy for cats. As their prime of life starts to fade, cats are considered to have reached middle age. Fading usually takes the form of a reduction in condition, or an increase in ailments.

Middle age

At about 7 years of age, a decline in the cat's general robustness may begin. It is the start of a phase in the cat's life that could be regarded as middle age, when it is less able-bodied. This stage continues throughout its years and into the double digits, i.e. when a full decade of its life has been attained. It is the cat's birthright to experience this phase of life, which it has always been fully capable of reaching, but was formerly too often denied. In times long past, due to indifferent attitudes and poor care by society, it was a fact that none but a minority of the general cat population lived for 10 years. Similarly, human longevity did not fare well either a few centuries ago. Through civilization's progress, times have changed immensely for both species. With regard to our cats, such progress has led to the ultimate objective of well-being.

With the continuing quarantined status of the house-bound cat, there is great assurance that its health will not face challenges from an outside source. Inner problems, however, can become emergent. For some middle-aged cats, the wear and tear of life can start to affect organs and tissues, while entirely new pathologies can develop in the animal's ageing body. Previously hidden disorders can now take advantage of any decline in the cat's state of health, becoming an established ailment. Notice must be taken of any sinister changes in breathing, chewing, eating, drinking, moving or excreting. Special notice of urination should be observed, and any sign of difficulty in passing urine should be regarded as a need for emergency veterinary treatment.

The enumeration of other specific disorders that could, in theory, affect a middle-aged cat would require a major textbook on veterinary pathology. What is relevant here is the standard of monitoring and care required for the ageing cat, and the caregiver's awareness of such ageing in a pet that may not exhibit signs of physical change. Cats can become highly inexpressive when they are developing an illness. When a suspected abnormality is detected and then confirmed as a disturbance to the cat's well-being, clinical attention is in prompt need.

Feeding calls for additional attention in this age group. Any free supply of food, for example, can be gradually altered to a given daily ration, of an amount previously determined adequate for the new age of the particular cat. Over-eating is to be avoided and a reduction of protein in the diet can ease the strain on the ageing kidneys. These measures aid in the preparation for the arrival of old age. Hairballs, anal glands, teeth, kidneys, allergies, parasites and various other items can become sources of health problems. For the outdoor-bound, middle-aged cat, the world continues to present a further source of hazards, including misadventure, trauma, infection and poison. With this are added the possible ailments of the middle-aged cat at home, to which allusion has been made.

Old age

Any cat that is 11 to 12 years of age should be regarded as having entered its old age. It may still enjoy health and well-being in this advanced level of feline existence if it has triumphed over any illnesses or physical disturbances, and remains in good care. In various reports, the normal lifespan of a cat, properly cared for, has been estimated to lie between 12 and 15 years of age. As already stated, 12 years was found to be the average life expectancy among apartment-kept cats in one North American study over 20 years ago. By considering other, more recent international estimates however, the cat's lifespan now appears commonly to be up to 15 years. This represents the phase of old age, as presented here.

During its old age, the cat can benefit from a degree of intensive animal care with freedom of mobility (Table 4.1). This means monitoring what it consumes and passes on a daily basis. Maintaining a clean litter box, grooming, petting, providing clean food and water, as well as a comfortable bed and a warm retreat are necessary aspects of care. Such intensive care is essentially the usual care, but practised more intently and with further attention to detail. For example, if the animal's appetite is failing, tiny portions of favoured food can be given. Matted hair would demand professional trimming and grooming, while constipation would require a mild laxative and sores may need antiseptic treatment. Not least of all considerations in extra special care is the old cat's need of sunshine for its well-being. To facilitate this, a comfortable resting place should be allocated in a window with sunlight.

As with any age, clinical attention should be provided whenever there is doubt about health. At such a time, special food and dietary supplements may be prescribed, in addition to any specific treatment that is warranted. The veterinarian might alert the cat's owner of any sinister challenge to their pet's well-being that may have arisen. A course of action could then be determined for the animal, taking into account any potential suffering.

Extreme age

The owner of a cat in this stage must consciously accept the reality that their animal is approaching the end of its life. It is extremely difficult and perhaps even impossible for every cat owner to be fully prepared emotionally for the death of their pet. Human–animal bonds can be very strong when the animal has shared the home and provided lifelong companionship in return for the fond care given by its household. At the time of pet loss there is much grief, which, regardless of the cause, has little ease from any rationalization. Grief is often expressed through caregivers' private remembrance of the intangible, vibrant qualities of their own, privately attached animal.

A cat that reaches its sixteenth year of life is considered a true survivor. It is surprising, nevertheless, how many cats do survive into this late and final period of life. In exceptional cases, cats can reach 20 years of age or more. Cat owners should not expect this unusual age for their animals. Sixteen years and over is equivalent to 90 to 100 years in human life. Through a combination of exceptional care, extraordinary good fortune, special genes and the full use of their legendary nine lives, cats can and do reach extreme age. Table 4.2 gives an outline of the special features of management for the different age groups of cats that are confined at home.

The valuable survey of cat features performed by the Feline Advisory Bureau and reported by Bessant (2004a) revealed, among many other facts, that cats frequently change in personality in step with ageing. Two-thirds of cat owners reported that the personality of their cats had changed with time and that the changes were more marked in older cats.

Table 4.1. Behavioural properties and freedoms in relative well-being in older cats.

General behavioural properties	Specific freedoms
Adequate physical mobility	Free to occupy permitted locations of choice
Motivation and ability to perform principal self-maintenance activities	Free from evident pain, disease, injury, discomfort or soiling
Periodic readiness and capacity to have social interaction	Free from fear, psychological distress or a lasting depressed state
Normal ability to ingest and excrete regularly	Free from hunger, thirst and retention

Table 4.2. Age groups and well-being management in house-confined urban cats.

Age group	Approximate span of age group	Wellness management procedures
Infancy	2 days up to 3 months	Successful suckling, human contact by frequent handling, progressive weaning by increasing supply of solid food, litter training
Juvenility	3+ months up to 8 months	Final weaning, good adoption, proper food, grooming, toy supply, parasite therapy, handling, vaccination, neutering, identification
Youth	8+ months up to 1 year	Containment at home, training to use scratching post and avoid plants, etc., worming, proper feeding, provoking play, social contact
Maturity	1 year up to 7 full years	Secure accommodation, own places, occasional play, grooming, good diet, parasite control, preventive care in general, social contact, checking teeth
Middle age	8 years up to 11 full years	Care as above for mature age group, plus extra grooming, health monitoring, petting, teeth care, physical checking, vet attention
Old age	12 years up to 16 full years	Care as above for middle age, plus daily monitoring, gradually altered diet, coat care, clinical visits as indicated
Extreme age	17+ years	Intensive application of care procedures as given above for the previous age groups, plus gentle handling and anticipation of euthanasia

Table 4.3. Feline personality change and ageing.

Personality change	Number	Percentage[a]
More confident	596	32
More interactive	502	27
Less nervous	416	22
More demanding	367	20

[a]Out of 1853 cats in total.

In age groups, it was reported that cats about 10 and 11 years of age were 1.7 times more likely to have changed in personality and cats over 14 years of age were twice as likely to have shown a personality change. In rank order, the changes reported were as in Table 4.3.

Further observations on changes occurring with age were as follows:

- Many other cats became more talkative, or more aloof, or bad tempered, or less bad tempered with age.
- Early socialization is of enormous importance in shaping the kitten to become an amiable cat.
- Cats often become more dependent on their owners as they age and can become over-bonded.
- Cats can become more fearful of travel and changes in their environment as they get older.
- All cats become decreasingly active by steps as they age, being noticeable after 2 years old.

Notable physical changes also occur with the ageing process, as there is deterioration in the body's fitness towards old age. (In the Feline Advisory Bureau survey, age 23 years in a cat is taken as the equivalent of a human centenarian.) Aged cats can develop unusual behaviour such as frequent confusion and night howling due to mental deterioration in a condition termed feline cognitive dysfunction.

5 Basic Activities

Taking the cat's internal physiology as a given factor, the domestic cat's life hinges on the intake of air, fluid and nutritional substance with the main output as the behaviour of self-maintenance. The most basic activities are: (i) ingestion and elimination; (ii) physical and social; and (iii) body care, exploration and rest. More comprehensive activities are given in Table 5.1.

Ingestion and Elimination

Water control

Cats are not particularly thirsty animals; their ancestors in the wild would have had limited sources of water, but they undoubtedly got some moisture out of the fresh tissues of prey. Even now, when they get a catch, they typically eviscerate the prey, consuming moist entrails. Cats drink sparingly by careful, rapid tongue lapping. They lap three to five times before each swallow. Outdoors, they may lick moist grass or smooth, dew-wetted surfaces or drink from small pools of rainwater.

Since cats do not sweat detectably and only lose fluid via nasal breathing and keeping a moist nosepad, they are experts at body fluid conservation. Physiologic fluid maintenance is vital for kidney health in the face of this cat characteristic. For this reason, housecats must have access to fresh, potable water at all times, to meet their modest thirst whenever this might occur, as it typically does suddenly and briefly. A supply of fresh, moist food is also helpful. Whether it is good for them or not, many cats have a predilection for cow's milk (with its minerals). This is usually a carry-over from their kittenhood and being fed such milk to wean them from their mothers. If they are not reared this way, they have little desire for milk in their adulthood.

Cats are, in fact, moisture misers and this explains so much about them. Although they open their mouths for their vowel-rich vocalizations, they keep their mouths shut most of the time. Until they are hit with episodes of intense play, they are not unnecessarily energetic. The physiological fact is that the more active the cat is, the more it will ventilate by breathing and the more it ventilates, the greater is its output of water vapour and this will cause a thirst. A thirst was inconvenient, even dangerous, in the Libyan Desert, which is still their 'old country'. They remain full of 'old country' genes.

Even in breathing, cats employ a tactic of water control. Some water vapour is transported and lost with each expiration, especially during oral breathing with the mouth wide open, as in panting. Breathing out via the nose allows the nasal membranes to re-absorb some vapour. Cats do not pant like dogs, with the mouth open and the tongue hanging out. They seldom allow themselves to become breathless, even if they might breathe rapidly and shallowly in some situations. The tear glands keep the large eyeballs moist. The tear fluid passes down ducts to the nose and this keeps the nasal passages functional. Their tight nostrils filter each outward breath, conserving moisture and defying evaporation.

When a cat does open its mouth wide and emits any extravagantly moist, oral sound, such as hissing or spitting, it is a sign that it is throwing its normal physiological caution and restraint to the wind and is ready to enact very serious physical business. Such a threat is not lost on a feline adversary, or any knowledgeable alien. The cat's major ethos of repose and inertia is not idleness; it happens to be its natural fluid-conserving strategy, one that would have been ideal for desert life.

Eating

When it comes to their tastes in eating, cats don't always stick to the rules. Here's one that likes tuna, but doesn't like salmon. Another won't touch fish,

Table 5.1. Comprehensive activities in the functions of cat well-being.

Activity	Forms	Functions
Reaction	Reflexes, responses in general, 'fight or flight'	Defence, avoidance
Ingestion	Eating, drinking	Nutrition, growth
Body care	Licking, scratching, nibbling	Hygiene, protection
Motion	Running, stretching, play	Exercise, energizing, practice
Rest	Sleep, drowsing, perching	Restoration, conservation
Association	Affiliation, bonding, social exchange	Companionship, alliance
Exploration	Seeking, investigating	Hunting, learning
Territoriality	Adopting special places, selecting own locations	Security, privacy, hiding

but loves chicken. One may love dry food, while the next prefers wet food. One has a taste for houseplants, another likes butter, one chews grass and the next chews wool. One likes to lick your hand, but another will bite it, or your nose, in evident affection. In fact, cats do have a major rule, which is that they don't always do the right or nicest thing (from the human point of view). This is a feature of the cat's culture. Perhaps this stems from that small percentage of the cat's nature that never took to domestication and resents it slightly. Cats can develop anorexia from taking a dislike to the food presented. In *ad libitum* feeding, the residual food can become stale, so the entire supply of food that is on offer should be regularly replaced with fresh food that has been properly stored where it cannot evaporate.

However, the cat is a classic carnivore and a very particular one. As previously stated, this is a very fastidious animal. A cat would never eat carrion, as some other carnivores readily do. The cat's meat has to be fresh. It eats only small portions at a time and never gulps its food. It does not eat or gnaw the left-over, mineral-laden bones of prey or bony kitchen scraps.

In proper care, a cat may eat several times daily. Many cats do much of their eating at night and this is consistent with their mealtimes in the wild. Their natural hunting times are at early night and dawn twilight, which are the times when small rodents come out and are abroad. They have extra-special vision in poor light and can focus their hunting attention very accurately with the help of their keen sense of hearing. They can also hunt by day and, for example, may catch birds at that time, using their special stealth and their high, bounding leap.

A system for feeding a household cat should be in accord with the animal's basic nature in order to cater to its orthodox appetite and digestion. The cat's natural appetite is usually well controlled. Obviously, this hunting creature's food should be fairly high in crude protein and very low in mineralized ash. Commercially available cat food has the proper proportions of these. Moist canned food has about 15% crude protein, 3% ash and 20% moisture. Some may contain much more water, even up to 75% of weight. Concentrated, dry cat foods have about twice as much protein and half as much moisture, about 10%. Both types contain sufficient fat, which is obviously part of the natural diet of prey. Commercially produced cat foods usually contain up to 20% crude fat.

By a process of trial and error, the exact quantity of well-balanced food needed by a cat can be determined by feeding the amount the cat wants daily, over a short period and watching the animal's condition. An exact assessment of its ongoing well-being would be even more valuable. If the cat gains too much weight on its given supply of food, some rationing would be required. In fact, the best way to keep the cat's weight in order is the provision of free exercise in a safe outdoor facility.

As a very general rule, about 100 g of food is needed daily for an average sized, full-grown cat (5.4–6.8 kg in weight) on moist food. For a cat of the same weight feeding on dry food, about 250 ml (a cupful) would be the daily requirement. Cats with unusual metabolism might need quite different amounts of food daily. Differing weights of cats obviously need differing amounts of food. Details of recommended quantity per weight are specified on the labels of commercially produced cat food. Mixed table scraps, if fed, should be chopped up since large portions of food are usually carried away for mastication at a further location from its dish and this distracts the cat's attention to its presented meal.

As essential carnivores, cats need to eat good quality protein. The stated quantity of protein in some commercial cat foods can be misleading, since some forms of protein are not digestible and therefore have no nutritional value. Digestible protein in the cat's diet can be ensured by periodically providing modest portions of cooked meat in such forms as chicken, ground beef or fish. This protein would be given as a supplement to the cat's staple diet without creating a major change in its rations. Varying such supplements can cater to the cat's sensitivity about food.

In some cases geriatric cats need a modified diet, perhaps one with less protein if kidney function may have become diminished in old age. Special diets are also prescribed for certain ailments. In all events, good nourishment is the principal basis of feline well-being and is therefore the most important aspect of cat care.

Drinking

Cat urine is very concentrated; as a result the cat retains as much body water as possible while excreting waste products via the kidneys. This physiological trick results in a reduced need for drinking water. With very moist food, containing three-quarters of its weight as water, the cat may not need to drink noticeable amounts of water from another source if it is healthy. Excessive drinking is not a healthy sign and warrants veterinary attention.

Cats drink by lapping gently with the sides and the tip of the tongue curled inwards and upward to form a spoon-like shape. The wet underside of the tongue in this shape draws up small quantities of fluid by flicking and sucking these, creating an upward flow of water into the interior of the mouth. While drinking, this animal is normally careful to keep its muzzle dry. Drinking periods are usually brief, about 10 s or less.

As a good moisture miser, cats do not drink great quantities. In fact, most housecats may seldom be seen drinking. Like many other creatures, cats also obtain water from an internal source, as a by-product of their metabolic processes. In spite of this, it is essential to provide a supply of potable water at all times to cover those important periods when the cat may need a drink. Most cats can take milk, but some are so badly affected by some of the constituents, such as lactose or some milk proteins, that they develop severe diarrhoea. The great loss of fluid from this, to an animal that lives by conserving its fluids, can be life-threatening. Small quantities of milk are safer, even for cats that are free of lactose intolerance.

Elimination

Cats are careful with their excretions and are innately 'toilet trained'. This is in keeping with their image of diligent self-care. Cleanliness vitally underpins wellness and this should be facilitated by cat caregivers following daily routines of hygienic attention to the litter box.

Urination

Both cat sexes urinate backwards with a fine stream. In the act of passing urine, they crouch down with hunched backs to direct the flow into the litter. To do this with some ease they need to have enough box space for this posture that ensures the hindlegs are kept dry and unsoiled, which is necessary for their comfort. The cat then turns around and scrapes litter over the wet spot. On average, cats urinate only two or three times daily.

Cats can use their urine for signage. They place territorial notices by spraying jets of urine on particular outdoor locations that are their boundaries, or favourite home sites, or places where other cats have urinated. They do this by backing up to the chosen spot with hindquarters elevated and the upright tail quivering, and then jetting the fine stream of urine backwards on to the target. Males, females and neuters perform this territorial spraying ritual in the same manner. When they are outdoors, tomcats are the most diligent sprayers; their urine has the strongest odour that enhances their identity (to other cats only) and emphasizes their message. On spraying, the efflux identifies the milieu as occupied by the sprayer. Cats have one major urinary problem. They are subject to kidney stones on their domesticated diets.

Defecation

Because they are moisture misers, adult cats usually have dry faeces. Fluid faeces are a sign of an illness of some kind. Moist faeces may indicate a digestive upset from unsuitable food. Intestinal problems can also result from pathological conditions, including infections resulting in diarrhoea. Certain diseases make their appearance by fluid or watery faeces and general ailing.

As with the cat's habit of scraping litter over the urinated spot, the same goes with defecation. The cat buries its faeces very deliberately in an eliminating ritual. An exception to this occurs with free-ranging tomcats. They deposit territorially and leave it uncovered as an obvious message for other cats to interpret. Breeding male cats communicate by their excretions very effectively, indicating their existence, their status, their location and function. This is done defiantly in the face of their controlled, domestic circumstances. Cats have a small fraction of their nature which is in non-compliance with some facets of domestication. The male's zealous concern with reproduction is an example of this.

Common Physical and Social Activities

Cats possess a lissom quality of progressive movement, being lithe, supple and agile in a state of well-being. Their gaits include creeping, walking, jogging and sprinting. The gait is the manner of locomotion and the stride is a single cycle of each leg in action, before the cycle is repeated.

Creeping

When they hunt in play or in reality, cats progress very slowly, stopping periodically as they creep along on their bellies. As they move, their forefeet are kept under their chins and their rear limbs are tightly folded under their hindquarters. The feet shuffle forwards in very short steps. If the cat happens to be hunting actual prey in cover, such as long grass or undergrowth, the creeping may halt and the cat will remain motionless with the head down and the ears erect, as it focuses on the prey. If the prey is in reach, the cat then makes a sudden, bounding leap, with the ears then forwardly directed. In this bound the claws of the forefeet are fully extruded for the catch, while the hindfeet remain padded for the propulsion and spring in the lift-off.

Walking

In the walk, the cat is actually marching in a brisk, four-beat stride. It lifts its feet cleanly with the claws withdrawn. The motion in this stride is started with both hindlegs pushing forward. The first step is made with one of the forelegs. After this, the hindleg of the opposite side steps forward. The remaining foreleg then goes ahead and the stride is completed with the step of the remaining hindleg. During the walk, the tail is extended straight and it is a healthy sign if it is kept firmly upright.

Cats meander outdoors at a walk and many accidentally displaced cats are known to have walked 100 miles or more, in finding their way home. Cats are certainly capable of walking great distances with this steady, even stride with its economical pace. Their thick footpads help in the many walking feats that have been recorded. It almost goes without saying that their sense of direction can be quite phenomenal. This fact is currently beyond comprehension.

Jogging

Cats jog in the same way that we walk. As we know, our walk is a two-beat stride that goes one-two, one-two, left-right, left-right and we swing the left arm forward when the right leg goes forward; similarly, the cat puts its left foreleg forward when the right hindleg goes forward, and so on. They can jog along at a good, steady speed in an easy manner using this two-beat gait, which is basically a slow trot. The cat's trot is a bouncing style of motion. In this gait, the cat uses one foreleg in exact unison with the diagonally opposite hindleg. The other diagonal pair of foreleg and hindleg then moves in step to complete the stride. This is the cat's hurried, but restrained form of running. The fit cat can jog for considerable distances.

Sprinting

In its sprinting gait, the cat leaps forward quickly on to the extended forelegs and draws its body and hindlegs forward. It extends the hindlegs forward as far as possible and plants them down to take the weight and push the animal forward, in a very outstretched style. The leading forelegs throw the cat upwards off the ground in a flying phase to let the other legs get underneath in the next furious stride. Since their bodies are relatively long and their legs relatively short, the hindlegs do not catch up with the forelegs in the sprint and the cat therefore has an outstretched style of running at top speed. While running at its fastest speed (galloping) the cat progresses by a series of leaps in rapid succession. In this gait, both hindlegs act in precise unison, giving maximal thrust to each forward stride by rhythmic extension and flexion.

While the cat in full flight uses the sprint, it can sustain this only for a limited distance, which may be no more than 30 m. If the animal nears its sprinting limit and is still in flight, it will usually try to leap upwards for elevated security, or dive downwards into some deep cover. Its course of action depends on what scope for escape there might be. Many cats go into a sprint spontaneously in an apparently playful setting. One example of this is the habit of many housecats to sprint around their accommodation in the early hours of night.

Climbing

It is well known that cats have great ability in climbing trees and their claws serve them well in the ascent. Coming back down is a very different matter. The claws are of no use then. This can cause an inexperienced cat on its first climb to find itself trapped out on a limb of the tree and unable to descend. Experienced, strong cats can deal with such a situation by crawling down branches as far as possible and then, when on the main trunk, they reach down to a lower point with the forelegs and from this point a spring can be made outwards, away from the trunk, using the hindlegs. The cat is then able to land evenly on the ground, with all four legs taking the impact simultaneously.

Jumping

In addition to the extensor musculature of the hindquarters and hindlegs, the cat's capacity for spectacular jumping has great support from the very flexible spine. The cat has a uniquely flexible spinal column that can be bent into distinct curves. A notable form of spinal flexion curves the back upwards into an arch. This is effected by contraction of the psoas muscles and other back flexors. With this type of posture, sudden spinal extension by contraction of the longissimus dorsi creates a forceful spring for the body. Many athletic cats can jump more than 2 m in length and over a similar distance in height. Jumping ability is essential in the feline type of predation.

Swimming

All cats are capable of swimming. Wild species, as those described, do so quite readily, while others such as the domestic cat would be reluctant unless a lifesaving attempt was necessary. Even young kittens a few weeks old are able to swim. Cats, when swimming, use the dog-paddle motion. Like running, swimming requires a lot of energy and effort, and the cat cannot sustain this exertion for very long. Exhaustion from swimming will ultimately lead to drowning if the animal is unable to reach ground. The antiquated method of eliminating unwanted kittens by drowning was an inhumane, cruel practice. Such an act would warrant the perpetrator a charge of animal cruelty.

Perambulating

Cats that are allowed to go outdoors can sometimes 'go walkabout', to use an Australian idiom. They may, of their own volition, leave their territorial core area and deliberately wander away from home. Do cats suffer occasionally from an innate wanderlust? Do they sometimes seek a new, different environment? Do they go to unexplored places simply for the experience? Do they pursue the walkabout in search of a mate or other cat company? Perhaps the answer is yes to all these questions; but the phenomenon – although real enough – cannot be totally understood unless an explanation can be found in the normal behaviour of wild cats.

In the course of its wandering it is easy to imagine what may befall the animal. The cat is likely to become lost, in spite of its long-distance homing ability. Perhaps in its walkabout it lacks the motivation of a cat removed from home unwillingly. The unneutered male, or tomcat, is the one individual with a territorial need. Because of its breeding drive it is also the cat most likely to become strayed, lost or road-killed. Cat road-kills in town or city environments are all too common and do not usually require to be reported. Live, stray cats are continuously collected by animal control authorities in every major community. They are then held briefly in pounds or shelters, awaiting destruction, unless claimed in time by their owners.

The lost cat in a suburban or rural location may find its way home after a while, or it may, on rare occasions, be taken into a new home. Again, there can be another strange result, namely the chance of a roaming cat joining a feline commune. Cat communal gatherings can become established in abandoned buildings in city centres, with the members living wild by scavenging. Feral cat colonies of considerable size exist in many cities. One of the more famous is in Rome, where citizens visit the colony daily to provide food for the cats.

The system of colony formation, both domestic and feral, is examined as a common problem for civic authorities. It is a prominent case for welfare intervention. The cat 'walkabout' phenomenon contributes to the outstanding animal welfare problem which is the annual, international killing of millions of un-homed cats by civic regulations. This matter should be addressed by society, which must accept that the cat-wandering-away problem (or phenomenon) now needs fuller understanding and wider recognition. As the major crisis in animal care, a solution is required. Since it leads to the condoned destruction of countless, healthy, very sentient pet animals, it makes other high-profile controversies in animal welfare look modest by comparison. Neutering and eventual re-homing by active public relations work might emerge as a solution.

The above problem does not, of course, usually affect cats that are permanently kept indoors, although such cats may inadvertently escape to the street. Indoor cats avoid the misadventures of those given daily free-range in urban environments. The expected lifespans of house-confined cats versus daily-freed cats in urban environments have been calculated in a North American study. Indoor city cats had an average life expectancy of 12 years, while those allowed outdoors regularly in cities had an average life expectancy of only 2 years.

Cats are innately leisurely and indoor cats can have a full and healthy life, given a few conditions. If they are neutered, properly fed and groomed, provided with good quarters, given veterinary attention as necessary and in receipt of abundant, caring human contact, they can enjoy the best of lives. The necessary exercise can be obtained by many normal activities such as jumping on to ledges, counters, shelves or any high surface, climbing stairs or playing with cat toys such as a table tennis ball. Activity can be produced in a cat if it is provoked into such activities as chasing, catching, fighting or hunting by the caregiver using stimulating lures. It is best to induce such action in young cats, leaving older cats mostly to their own devices. Spatial satisfaction can be obtained by cats that have access to a place that provides a good visual field, such as a wide window ledge or a table by a window with a broad, open outlook.

Use of the extensor system

In the frequent outstretching of its limbs, the cat is exercising its extensor muscles (Fig. 5.1). These

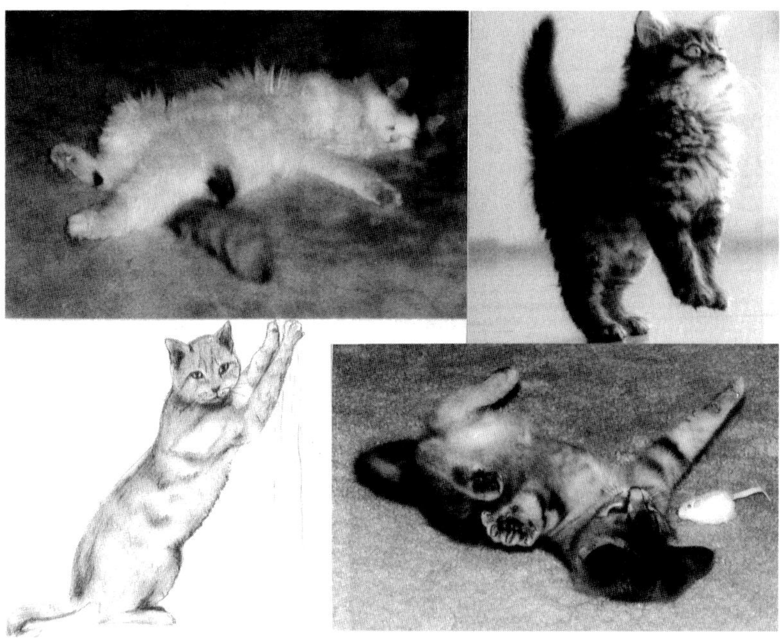

Fig. 5.1. Extensor system in use.

muscles constitute the system of muscling needed to propel the animal in its forward and upward leaps. Such movements are in the basic feline activities of pouncing, climbing and jumping, which are components of hunting (Turner and Meister, 1988). The muscling involved in these activities requires to be strong and, in most cases, to function in synchrony as a whole system.

In the foreleg, the joints that require extensor muscles are the shoulder, elbow and carpus. The principal muscle for extension of the shoulder is the supraspinatus. The brachiocephalic and the triceps extend the elbow. In particular, the two shorter heads of the triceps (a broad three-headed muscle that originates in the scapula and is inserted in the humerus) effect much of the extension of the elbow. Together with the short anconeus muscle, the triceps is chiefly responsible for extension of the elbow to absorb the impact of landing. The common and the deep digital extensor muscles extend the carpus and the foot (Dyce *et al.*, 1996).

In the hindleg, the extensor muscles for the hip are the middle gluteal, the biceps femoris, the semitendinosus, the semimembranosus, the adductor and the quadratus femoris. The tensor fascia latae extend the stifle. The tarsus is extended by the superficial flexor and the foot is extended by the digital extensor muscles. The collection of extensor muscles is needed to absorb the impact when landing from any height that has been attained by the cat in the course of territorial surveillance. Since agile climbing is notable in the cat's repertoire of kinetics, the animal must be anatomically equipped for the corresponding descent. Landing from a height requires the bodily dynamics of synchronous and instantaneous action of all the components of the extensor system.

With its highly developed sense of equilibrium, the cat dropping from a significant height is quickly able to orient itself so that it lands evenly on all four feet. From lesser heights, the impact is taken by the forelegs and by the triceps muscle particularly. The main joints are extended strongly to counter their flexion. The maintenance of extensor function is provided by the high frequency of extensor exercises that healthy cats perform. For example, the forelegs are well exercised in stretching and the hindlegs receive their muscular fitness in jumping. These exercises of stretching and jumping also serve the orthopaedic needs for routine and effective movement in joints in this typically lithe animal.

With the high frequency of stretches and jumps in the healthy cat's behaviour, it is evident that such exercises are programmed into this animal's innate repertoire of diurnal activities. While housecats can perform their stretches easily on the floor, they require upper surfaces to land on for their jumps. The cat's household must have toleration and facilities for the animal's routine jumps. If, for any reason, such as an orthopaedic condition or inadequate accommodation, the cat is unable to perform frequent stretches and jumps its well-being is impaired.

Oddly, the feline extensor system appears to have greater force than the collective power of the flexor muscles. The greatest use of the latter is in running, when extension and flexion play equal parts in the energy expended (Fig. 5.2).

They can sprint briefly at speed, but cats soon lose their energy in running, as especially evidenced in the hunting behaviour of wild cats (Sunquist and Sunquist, 2002). For example, wild cats give up their attempt at capture, rather than continue the chase, if the prey escapes. However, if a capture is made, the cat has sufficient extensor strength remaining to drag heavy, dead prey over considerable distances before eating it. While the extensor versus flexor disparity may appear odd, it is fitted to the feline behaviour that features such characteristic leaps and bounds.

The longissimus dorsi is the largest and longest muscle in the cat's body, running the length of the back from the neck to the sacrum and is attached to each dorsal spinous process between these two points. Half of this muscle's mass is on each side of the spinal column and its origins and insertions are on all the vertebrae from the sacrum to the final cervical vertebrae. When both halves contract, there is powerful extension of the back and loins. When only one side contracts there is lateral flexion of the spine. The action of this muscle creates the oestrus posture of lordosis. This is therefore the muscle that gives the cat's back its remarkable and unique flexibility, given the capability for a degree of intervertebral mobility possessed by all members of the cat family.

Suppleness

Basic to the cat's characteristic suppleness of body is its spinal flexibility. The degree of flexion that its vertebral column possesses makes the fit cat uniquely acrobatic among all the domestic

Fig. 5.2. Feline sprinting; note the tiger's limbs in extension and flexion.

animals. Assuming its spine is its long axis, there is capability for sections of this vertebral shaft to have both rotation and flexion to varying degrees. It can twist or bend its back into remarkable movements and postures.

Rotation occurs in the way the cat can twist its body during a fall, in order to land on all four feet simultaneously. It can bend its back in the middle to turn its fore or hind end to either side, with great flexibility. In the 'fight or flight' display the back is convexly arched so high that the cat's size is greatly increased when viewed obliquely or from the side. In the female cat during her heat period, she displays her condition in a posture technically termed 'lordosis'. In this, the back is curved downwards in a concave way so severe that the cat's belly can touch the floor as it walks. Further, the cat's body can twist and curve vigorously in fighting or in struggling for any reason. Apart from the tail, the most flexible section of the cat's vertebral column is its neck. Such flexibility allows the animal fast direction and redirection with its head and mouth for focusing and biting. These capabilities are obviously great assets in hunting. When the cat is lying flat out and asleep, it sometimes has its head turned round 90° so that its face is uppermost in direction. This demonstrates the amount of flexibility that can be sustained in an unconscious state. With voluntary effort while awake, the cat can turn its head further.

The cat's tail is flexible throughout its length, being able to be twisted and turned at the top as strongly as at the middle or root. This is beyond a dog's ability. The cat uses its tail for such purposes as signalling its emotions and as a rudder when airborne in great leaps.

Playfulness

All the young domestic animals have their own ways of playing: lambs gambol in their fields, goat kids frolic with each other, puppies play with slippers, newborn calves do a little dance, foals race furiously. What is animal play? It is essentially an unnecessary activity performed vigorously with fast movements that involve most of the body. Clearly it is exercise that is innately wired into the animal's nervous system. Except for cats and dogs, it shows up most in the very young animal. Cats in particular are fond of playing throughout their lives.

In kittens, play begins at a very young age, becoming fiercer as they grow. It is not just fierce in form; it is also pursued with a fierce attitude which is directed at the plaything. The cat at play is obviously acting out its roles in the wild, with stalking, catching and killing prey; the proper, serious mood goes with the acting. Younger cats have an almost insatiable appetite for play, especially if the play period gets human aid. They most readily play with other cats in the form if fighting. They can play by themselves, of course, and the availability of cat toys can help them to direct the activity and get this behaviour out of their system. The most suitable toys are items that slightly resemble natural prey – small, mobile and perhaps soft or fluffy.

The cat gets good exercise in play, since it puts its best effort into it and the session represents a good, whole-body workout. When the workout has been adequate, the play ends. The performance also indicates the cat's general fitness and health, of which the caregiver should take note. The cat is likely to show some aggression in play, so it is advisable for caregivers to ensure they do not become identified as the play object. The aggressiveness might escalate as the cat matures and with the hard wiring for this activity in the cat's nervous system, the animal might get carried away with the game that it is performing instinctively. Unfortunate misunderstandings between cat and caregiver can develop, unwittingly. It is therefore necessary for the caregiver to understand the biology of cat play.

Stretching

From their ancient, prehistoric days, cats have retained their full genetic program for the hunting lifestyle by which they then survived. Much of their domestic play today is exercise in hunting surrogate items of prey, even their own tails, or phantoms of prey. A form of stationary exercise is a cat specialty, namely stretching.

In spite of its superficial image as a lethargic animal, the cat is a good athlete. It has great suppleness of body, speed of action, fast reaction, keen senses, gymnastic ability, precise motion and equilibrium. To maintain physical fitness the cat has an exercising discipline in the nature of aerobic action. This takes the form of a pattern of systematic stretching, homologous with the human stretching phenomenon, termed 'pandiculation' in medical terminology (Figs 5.3 and 5.4).

In humans and cats alike, pandiculation occurs as a conscious, yet involuntary outstretching exertion which sweeps over the subject's major mobile parts, from end to end. With pandiculation, there occurs a chain of full extensions involving the head and neck, the spine and the limbs. At times, while the cat is standing the general stretching includes the upper spine plus the fore parts only, or the lower spine and hindlimbs only; thus partial, or polar pandiculations exist. Minimally, either the cat's fore- or hindlimbs only may be outstretched.

Fig. 5.3. Full pandiculation in recumbency.

Fig. 5.4. Pandiculation on belly.

In the case of the forelimbs, they are usually stretched together with the forepaws on the ground. With hindlimb-only extensions, alternate legs are stretched back off the ground. Partial pandiculations are more common in older cats.

In pandiculation, the act of stretching is characteristically slow. This allows the muscles that are being extended to act passively. A rapid stretching action of a body part or limb causes counteracting/opposing muscles to stiffen reflexly and this prevents relaxation. Pandiculation assists in maintaining muscular well-being. The full extension of the joints that are included is very beneficial in maintaining their function also. Thus, pandiculation services the orthopaedic system overall, something vital for a hunting athlete such as our antiquated cat.

Young cats often perform full pandiculation while lying on one side. In this position the animal typically curves its whole spine down and vigorously extends its head and forelegs forwards and its hindlimbs backward. This display is one of remarkable strength, with all of the extensions occurring simultaneously and being sustained for several seconds.

Pandiculations, whether complete or partial, are performed by healthy cats numerous times daily, and they often follow periods of rest or sleep. On waking from sleep, cats often yawn widely and yawning in humans is formally considered to be a component part of pandiculation. Yawning serves to exercise the cat's jaw joints and muscles. Doubtless this maintains the cat's capability for biting and chewing, which are basic functions, vital to life in fact.

The effects of routine pandiculated stretchings are very significant and include maintaining the mechanics of movement in joints for their state of function, the toning of musculature for strength, and the improvement in blood circulation to the orthopaedic structures overall. It appears to be pleasurable to the subject, being performed readily and satisfactorily in healthy cats. It can therefore be taken as a reliable signal of well-being. As some evidence of this is the fact that this systematic stretching ceases if the animal is unwell in any way. While other animals also pandiculate, the cat is the principal performer of this phenomenon.

While occupied with resting, the cat is effectively killing time until the next sudden and extravagant outburst of energy for capturing prey. That was the main key to its survival in ancient days. This inherent restraint is an irony of ironies for the cat in maintaining its athletic fitness. However, since this conservation of movement in rest acts counter to the exercise requirement of this athletic animal, the innate regimen of systematic stretching takes over to meet this need. Pandiculation often occurs after a protracted period of sleep.

Yawning

Yawning often occurs as an integral component of the compound stretching exercise of pandiculation (*Dorland's Medical Dictionary*, 2003). This oral action in cats is usually performed during periods of inactivity without the stretching of other bodily parts. As a singular exercise, feline yawning is performed with particular vigour, causing the mouth to be opened so that the upper and lower jaws form a 90° angle of separation. This extent of mouth opening is not observed in most other animals, but is a characteristic of felines from the largest wild cats to the common housecat.

One function of yawning is to fully articulate the temporo-mandibular joints, stretching their related muscles. This exercise has the effect of maintaining optimal efficiency in mandibular joint function during periods of oral inactivity. The daily incidence of yawning in cats, both wild and domesticated, is higher than commonly occurs in other animals. Evidently, the capability of extreme jaw extension in felines is important to maintain (Fig. 5.5).

The relationship between wide yawning and vital life is apparent in the feline manner of predation. Most wild cats include in their prey animals that exceed themselves in size. After catching such prey, the feline method of administering the *coup de grâce*, or executing stroke, is by gripping the prey at the junction of its head and neck in a wide bite, throttling the animal and causing a prompt death. The encompassing bite requires an exceptionally wide expansion of the jaws. This is rooted in the feline genotype as a special piece of behaviour. The related yawning in a cat serves as an indication that the animal is in functional health or in a condition of well-being.

Body Care, Exploration and Rest

With this animal, so fastidious in its eating habits, and so meticulous in its personal hygiene, it is only appropriate for it to be exacting in its body care, which it is. The principal forms of this care are washing, grooming and avoiding contact with wetness or dirt.

In the practice of avoiding contamination cats take careful steps where the ground or floor is wet or soiled. They flick their paws when these get wet. They have very sensitive paw pads and make every effort to keep these free of any contaminating matter. They jump to avoid pools or patches of mud when walking outdoors and are averse to snow. They pick their way very carefully over any surface covered with rough stones or gravel. They prefer, whenever possible, to use a high, narrow level and smooth walking surface (a catwalk) rather than travel at ground level.

A notable feature of floor or ground level avoidance in cat behaviour is perching. They perch in a resting posture that resembles crouching, except that the forepaws are folded into the chest area when perching. The perching posture

Fig. 5.5. Feline yawns; note the jaw extensions.

is adopted in a resting phase (or simply an inactive one) during which the cat may drowse or remain awake and watchful, while located on an elevated level site indoors or outdoors. Within a house the perching place may be an arm of a couch or chair, an upper step of a stairway, or the flat surface of an item of tall furniture. When they are outdoors cats readily perch on a fence post, the flat rim of a railing or a wall, indeed the even surface of any tall structure tactically located as a lookout. The ultimate in perching is on the branch of an available tree that can be scaled. Healthy cats perch periodically each day and perching periods may last for long periods if undisturbed.

Washing and grooming

Cats have grooming sessions fitfully. The session may be a brief one of scratching or nibbling, or a more systematic one of washing. The cat is diligent in attending to its washing. It washes itself in a grooming activity that follows a set pattern. With its raspy tongue, it begins by licking the inner surface of the wrist and paw of one forefoot to create a washing sponge. The cat then sponges the side of its face and forehead on that side using forwardly directed, curving-over motions that sometimes take in the eye and ear on that side. The other side of the head is attended in the same manner by the opposite forefoot. The sponging parts of the forefeet are licked frequently during this washing exercise. In this activity, the cat can rotate its wrist in human fashion (Fig. 5.6).

If the cat is disturbed during its washing session, it will likely move to another location to complete the process which may last about half an hour.

Since the cat has a very flexible spine, it is able to sit back on its croup area and face its underside, namely its chest, abdomen and its thighs. In this position these parts are groomed by long licks that take the fully extruded tongue from the roots to the tips of the hairs. The hindquarters and the tail are similarly groomed from this special, laid-back position, which is not the normal sitting posture. Of course, the cat can lick its breast, shoulders and forelegs in the normal sitting position. Again, the flexible spine allows the cat to turn its neck and head to either side in order to groom its flanks by industrious licking.

Scratching rapidly and briefly is involved periodically in grooming whether or not the animal has fleas or skin irritants; however, severe and persistent scratching with self-biting is not simple grooming, but is probably a sign of parasitic infestation calling for veterinary attention. Cat dander is removed from the cat by scratching and also by rubbing the sides of the head and body against persons and structures in the cat's usual domain. Some of this dander can remain airborne for a while and can be allergenic in susceptible people.

By nibbling through their coats down to the skin, cats abrade small areas of the epidermis from time to time, removing dead skin cells that contribute to the formation of dander. Nibbling at their skin surfaces is a routine and occasional item of grooming in healthy cats and is usually directed at accessible areas of the trunk. As with scratching,

Fig. 5.6. Kitten preparing its forepaw to make a 'sponge' for face washing.

excessive nibbling is usually a sign of skin parasitism by fleas or lice. Infestation with such ectoparasites prohibits well-being through the irritant suffering which results from the pruritus that develops in response to infected flea saliva.

Evaporation control

In general grooming, while the cat is licking parts of its body, or an associate, the licking is essentially dry combing with the lingual barbs, or tines. These are abrasive, hooked, filiform papillae on the centre of the cat's tongue (Bradshaw, 1992). However, the cat loses saliva when it is repeatedly licking the dorsum of its paw and carpal area to wet it to form a sponge for washing its face. Such wet licking is the cat's only extravagance with its fluids.

In the cat, the sweat glands exist only around the sole of the foot and the interdigital skin. Evaporation of fluid can occur at that pedal area. Notably, in this fluid miser, the cat at rest limits evaporation from this sweating area by keeping its toes close together and the foot flexed. Similarly, when the cat is in a heated state, it readily allows its feet to become wet by dipping, or flicking, them in water. Heat can then be lost via that area.

Scent production

From above the eye and into the ear, on each side of a cat's head, is a broad patch of skin poorly covered with hair. These patches lie over the temples of the skull and carry masses of very small apocrine glands that secrete directly on the skin. The product of these specialized glands is a marking odour that has no smell to humans. Every type of cat and every feline species (and they alone) have this strange anatomical feature. It is not a useless vestige of some part of the body that only had a function long ago; it is a feline organ that still has some territorial purpose with its fine exudation. This secret odour is beyond our evaluation; it is obviously aimed at other cats. Adding to their odour-marking, cats have a pair of anal glands.

The secretion of the cat's temporal skin is believed to contain the odour of its identity. On some occasions – when perhaps it is only motivated by a whim – the cat rubs, or swipes, the zygomatic ridge of one temple on an object in its occupied territory. Apparently, this is its way of marking its possessions as a message to other cats. At times the cat may be seen to swipe the zygomatic ridge below the temple very firmly on some upright surface.

The swiping gesture should not be confused with the wiping action, in which the cat rubs the wet corner of its mouth on its human associate. This wiping gesture may be done to signal the feline version of platonic love, which has been mentioned in Chapter 1 of this book. On the other hand, it may be that swiping and wiping are simply two more methods in the cat's territorial marking repertoire, which includes scratching, defecating, spraying and flank rubbing. Every cat seems to be keen to leave its mark in life and to let other cats know that it exists.

Rest and sleep

The principal feature in the cat's manner of living – its *modus vivendi* – is its adoption of numerous phases of inactivity every day. Such phases may or may not be regarded as deliberate rest-taking sessions. 'Rest from what?' it could be asked. These sessions usually do not follow any energetic activity. This inactivity is in accord with the cat's characteristic habit of conservation in bodily action.

Cats choose various resting and sleeping places and occasionally seek a hideout in their homes. Unlike dogs, cats do not like to sleep at floor level, even if they are offered comfortable beds in baskets on the floor. They like to adopt sanctums and to occupy recesses of close containment as places for secreting themselves for undisturbed rest. The resting postures used while the cat is awake are all upright in form. Three of these occur as follows.

1. *Upright sitting*. The animal rests on its hindquarters and vertical forelegs with the head held upright.
2. *Sternal resting*. The animal lies on its belly and on the floor of its chest. The elbows are withdrawn while the forefeet and head are extended forwards.
3. *Perching*. This position is a modification of crouching, but is more gathered-up in form. Typically, the forefeet are folded up and drawn into the chest. The head is less extended. It is usually adopted on an elevated location. Drowsing may occur in the perch position. The perching posture is adopted in a resting phase, or simply an inactive one, during which the cat remains awake and watchful while located securely indoors or outdoors. Healthy cats perch periodically each day

Fig. 5.7. Circular sleeping posture; note the degree of spinal flexion.

Fig. 5.8. Sternal resting.

and perching periods may last for only a few minutes, or persist for much longer durations if there is a peaceful environment.

Sleep is often taken lying on one side. In this posture the cat lies flat out with the extremities extended. In this form of sleep the cat is easily alerted, showing wakefulness with open, watchful eyes and head movements in response to any minor stimulus of sound or movement. This is cat-napping.

The more definitive sleeping posture is one in which the cat lies on one side and curls up into a very exact circular shape. The back is bent tightly. The tail is drawn into the opposite outline of the circle and its tip goes to the head. The head is tilted to one side and may even be rotated upside down. It is drawn into the neck and the folded forelegs. The hindlegs are extended under the neck. The circular mass is compact and this causes the abdominal breathing to be noticeable. One ear is often cocked above the circular mass or outside it (Fig. 5.7).

The cat has deep sleep usually in the circular position, from which it does not wake too readily. During this sleep the cat usually makes a few shifts in its position, but sleep continues during these small movements. Sleep in the circular posture commonly lasts half an hour, more or less, and it can be repeated several times, soon after each one of its sessions, with wide yawning between these. The respiratory rate drops slightly from the norm of 30 beats per minute during this sleep. The amount of sleep that cats take each 24 h period is commonly given as averaging between 18 and 20 h. The total time of relative inactivity is usually about that time, but this quantity is composed of the three conditions of resting, light sleeping and deeper sleep. More than half of the inactive time is spent in true sleep, while resting occupies the remaining time. Simple resting is a major occupation for cats. They may rest in unusual postures for brief periods, but resting on the sternum is the principal posture for this form of inactivity (Fig. 5.8).

It appears that cats dream. Occasionally in sleep they show all the signs that are taken to indicate dreaming. At such time there are slight twitchings of muscles, paws or legs. In association, there may be some jerking of the ears or parts of the body. These signs may be quickly repeated, but are not maintained for long. The short episodes of limb jerks and muscular twitching during sleep are biologically acceptable evidence of dreaming that is the equivalent of rapid eye movement.

6 Association and Reproduction

Association

Of all the dynamical arrangements for an indoor cat, one of the best is a compatible live companion, such as another cat. Even a dog or a pet rabbit can fill the need for animal company and cats are known to have bonded well with various other small livestock. The bonding process develops best when the associates get to know each other during the sensitive period of youth. Such live associates are of particular merit where no one is home with the cat during work times. When there is a bond between two cats, it is common for them to groom one another from time to time and occasionally they may kiss each other by pressing their muzzles together for a brief moment. Many single cats left alone for an unusually long period of time show a disaffected attitude to their caregivers when they return. This can endure for several hours. It is a minor feature of pure cattiness.

The pairing of cats may have initial difficulty. Depending on its nature, a resident cat can sometimes be resentful, even hostile, to another cat on account of implicit rivalry. Creating a trio can be even more difficult than pairing. This is particularly likely if there is an established twosome of cats and a third is added. One of the original pair may become socially intolerant in such undesirable ways as spraying in the house. Jealousy may be at work; even after the newcomer appears to have been accepted, anomalous behaviour may become shown by an earlier resident cat. Perhaps cat jealousy can stem from the natural, feline trait of being suspiciously vigilant.

Any number of cats beyond three is a colony and colonial behaviour in cats is essentially harmonious. Colonies of several dozen cats can sometimes develop in a home from uncontrolled breeding. It results from unrealistic ownership. Eventually this type of situation becomes a civic problem, arising from neighbourhood complaints. The matter is ultimately resolved by the intervention of local authorities, including veterinary services and often the police with enforcement documents. In these large house-bound colonies there is an absence of cat well-being to such a degree that mass euthanasia seems warranted. The event is always one of extreme distress to the cat owners and this may call for professional counselling.

The free-range feral colonies can become similar problems and real veterinary expertise is essential in resolving the situation. The colony may have become excessively large, it may have become a safety problem, it may harbour feline disease, or it may have become a hazard to human health. Determinations of health may be performed on captured or tranquilized cats to act as samples of the whole colony. This can be a first step in addressing the problem. Mass neutering of healthy individuals in reduced numbers could be a humane course of action, but may be an unrealistic one. Public relations work could determine an acceptable course of action. A humane course would require public approval, adequate funding and organized management of the mission's logistics. The only alternative to this is the same as for a house-bound colony, namely euthanasia. Full wellness is unlikely to exist in a dense cat colony. Density occurs when there is more than one cat per 50 square feet of enclosure space, as observed by experience of laboratory cat colonies.

Although we use the word 'domesticated' as a blanket term to cover those species of animals that have become allied to us, there are two levels of that alliance. The food animals are enclosed on the farm, since, in reality, they are only semi-domesticated. The well-domesticated ones are in our homes by mutual choice. Cats and dogs clearly have some innate desire for association, plus a desire for food and shelter, of course. Even alien relationships will suffice in meeting their social need. We meet their social desires with ours, to the material and psychological benefit of both;

the provision of food and shelter seals the deal for the animals. Cats have suppressed their vestigial, wild feline factor in the domestication deal. Reference has already been made to this hidden factor; it is still there in some cats, maybe in many and perhaps in all.

The cat's associative need is tactile more than vocal. Actions that apparently provide assurance and pleasure to a pet cat include stroking the head, sweeping strokes along the back and rubbing the temples. The human caregiver has to keep the association real, constant and effective for its maintenance in the cat. Poor or inferior associative care from the human party may terminate the emotive component of the association, if it was ever established. Perhaps the walkabout in some cases is a manifestation of such deterioration, or termination of the association.

In the cat, the human–animal association is not always as durable as in the dog. One advantage of this is the cat's greater adaptability to being re-homed after a displacement. On being re-homed, the cat is capable of developing a new, bonded association. In many cases, with good upkeep, the state of the cat–human association becomes a strong mutual bond. The association may also be a preferential one, some cats showing a preference for either female or male company. Housecats in general are very tolerant of children, especially if, in early life, they had familiarity with youngsters. Cats readily connect with childlike, behavioural characteristics.

With house-confined cats that have only one or two owners, the extra company of visitors usually animates the animal. At such a time, the cat typically makes physical, social contact with the extra individuals in friendly fashion, particularly if the friendly contact is promptly reciprocated. The Siamese can be exceptions to this rule.

Houses with sufficient space can contain half a dozen cats or more if the owner is cat-wise and knows all about multiple-cat husbandry. The commitment in cost and care is considerable in such cases and the arrangement is not normally advisable. However, the cats can provide ample association with each other. They develop their own relationships and overall social structure within the group. Unless they are all neutered, the size of such a domestic colony cannot be controlled and there is the danger of the pet colony becoming a mass cattery of the type that creates a civic problem, of the type already described. Often in these dense groups diseases are common, particularly feline leukaemia and feline AIDS. Efficient body care is difficult to maintain in multi-cat groups.

Interaction

The incidental behaviour of many cats in social situations can include negatively directed acts such as aggression, avoidance or disengagement, while activities of affiliation involve a positive relationship between the subjects. Such interactions of affiliation include rubbing the head or body on the partner, head-butting the same, allogrooming and lying in close physical with the associate. Minor vocalizations often precede interactions. Such acts are of two types, one being attention-seeking, the other is attention-giving. Examples of the former include head-butting, vocalizing and squatting, while examples of the latter include allogrooming and lap-sitting.

Reaction

The cat's character is typified by the type of reactivity it usually shows to stimulation. When an animal becomes accustomed to some stimuli it may lose its responsiveness. In general, cats keep their innate reactive character despite experience. The character, or personality, of the individual remains fairly typical of the individual but can suddenly alter if the cat is seriously sick. In the cat population there is a broad spectrum of reactivity. This is in accord with the first rule about cat character, which is that while all cats are truly feline in their ways of living, every cat has its individuality of character.

In reactive character, a minority are almost unreactive, showing little responsiveness when handled or manipulated. At the other extreme are those cats that react aversely to casual handling such as lifting. Within its individuality there may be degrees of change with age but the character core remains unchanged throughout the ageing process. For example, the great majority of kittens that react readily to sudden or unusual stimuli will keep much of that characteristic throughout a life of well-being.

Cat handlers should be aware of a risk in a particular reflex. In stroking a highly reactive type of cat, an aversive reaction may occur in it. With the cat lying on its side, hand pressure along its abdomen can stimulate the prey-killing reflex. In this, the cat encloses the hand, grips it with its forelegs, bites it, and vigorously tries to scratch it rapidly and repeatedly with the hind feet acting together. This is the usual method of killing prey, but less reactive cats will only use it in reality.

Cats are inquisitive by nature and are very reactive. That cats are highly motivated to investigate is very well known. 'Curiosity killed the cat' is the maxim that tells us that cats take exploration to dangerous limits such is their strong inborn drive to scrutinize things in their immediate environment. Their investigative method is visual inspection over small items. If the item drawing the cat's attention has movement the animal becomes instantly active in a positive reflex attempt to close in on it; for example, a cat will spring at a fly or a mouse or a bird or a butterfly with equal zeal. They often make close inspection to the caregiver's activities.

To sudden loud sound, the cat's usual reflex act is negative, taking instant flight or quick withdrawal. Most cats do not ever become totally accustomed to certain sudden or loud noises such as door bangs or vacuum cleaners, in spite of periodic exposure to them. Harsh noises do not contribute to a housecat's comfort or well-being. An unsettled cat may be liable to wander away in the fashion already described.

A notable cat reaction, which is neither positive nor negative (but perhaps both), is the state of 'fight or flight'. In this classic posture, the cat has a highly arched back, the eyes are wide apart and the pupils are dilated, the legs are straight, the tail is held down stiffly and all the hair is erect. The animal stands obliquely to the stimulus. After being maintained for a period of several seconds, the state may be resolved by changing to aggression and attack, or, more usually, by slow dissolution and withdrawal. The manner of resolution is apparently dependent on the cat's appraisal of the threat and the stimulus and its own underlying hormonal content.

Cats display other reactions and intentions through body language and facial expression. The great German ethologist Professor Paul Leyhausen (1979) studied facial expressions in cats and was the first to show that these were meaningful. Today, many recognize in cats that averted eyes display avoidance, a fixed stare indicates a social connection or singular attention, i.e. curiosity. When the whiskers are directed forward it indicates that the cat is switched on for aggressive action. Ears pricked forward spell confident playfulness. Fear is shown when the ears are drawn back; but anger is expressed with the ears folded back very firmly (for protection if a fight should ensue). Half-closed indicates satisfied relaxation or contentment with the body in the crouching position. With ears folded back to their limit, the eyes staring, pupils dilated, the mouth open and hissing and whiskers bristling forward, the cat is showing its most aggressive reaction to a perceived opponent and its intent to attack.

In rare cases a cat may become habitually and unacceptably aggressive towards the caregiver or other innocent persons. They are usually mature cats and the cause of the underlying pathological condition is usually unknown. Of course the aggression may have progressed by habit if tolerated. Unfortunately, a successful treatment may not be found. It could be speculated that the wildness factor inherent in the feline genetic infrastructure to which reference has already been made, although suppressed in domestication, can emerge like a mutation in rare individuals to create abnormal aggressiveness. Unless some treatable cause is discovered, incurable individuals require relief from the apparently psychotic state in euthanasia.

Again, on very rare occasions, a queen cat that has freshly delivered kittens may show uncontrolled aversion to them. She may bite them, eat parts of them and may also show aggression to the caregiver trying to interfere. The cause of this psychotic behaviour is not known but a cure can often be effected by administering a general anaesthetic to the cat so that it remains unconscious for a period and permitting it to recover in its own time. On recovery, the cat is likely to have normal maternal behaviour.

Reactive, noisy fighting between cats is likely to be no more than a natural quarrel or an effort to resolve disputed status. Real injuries, however, can result from the vicious bites and scratches; these may require first aid. The fight ends when one submits reactively.

Lying on the back with the belly up can occur as a reaction to provocation in playful fighting with a sibling, a companion cat, or even a caregiver. It is play defence, not true submission. Such a position when faced with a real aggressor would facilitate biting and tearing with the claws of the stronger hindlegs.

Reproduction

Unless it has been spayed or neutered, the kitten generally reaches puberty between 6 and 10 months of age. From then on it is very involved in breeding. On behalf of the dependant creatures in their care, cat owners must make a prompt decision on this important matter as soon as the animal is acquired. Although breeding is a major activity among cats, it is a hazardous business for them. Of the things that can go wrong in the pursuit of this activity, common ones include road-kills, loss, severe bites, disease transmission and unwanted kittens destined for destruction. The answer for such events is the excision of the gonads – the testes or ovaries – before puberty. In theory, house arrest can prevent the cat from mating, but in practice this fails too often. Through the hormone-driven determination of the breeding cat to escape from home in order to mate, its quick dart outdoors ruins the cat owner's good intention.

In the case of intact female cats successfully confined at home, the long phase of the animal's oestrus puts a demand on the caregiver's attention. The frequent recurrences of these futile heats eventually strain the human–cat bond and do nothing for cat wellness. In the matter of family planning for cats in general, the optimal number of pregnancies is zero.

When cat breeding is intended, the dynamic and physiological features of feline reproduction require to be appreciated by the cat managers. In the series of events completing the reproductive process, the principal ones are obviously oestrus, mating, pregnancy, birth and kitten development.

Oestrus

As they come in heat, cats get into a tangle of actions that make them look like they are suffering. The normally reserved female cat undergoes a radical change in behaviour when she is in oestrus. The contrast with her previous conduct is startling to the inexperienced owner. When they first observe their cat in heat, novice owners are often convinced that the animal is suffering from a painful disorder such as colic.

Between spells of normal-looking behaviour, the oestrous cat goes into gymnastic displays that include body rotation, creeping, belly-crawling, elevating the hindquarters, dipping down the spine, arching the tail and pressing the chest to the floor. Vocalizations that are associated with the display are typically harsh, unpleasant, crying sounds of howling or yelling, quite unlike the euphonic, gentle tones of meowing or mewing in normal cat calls. Of course, cat-to-cat communications are sometimes in a different language from that used when addressing humans. It could be argued that cats have both polite and impolite vocabularies.

In female cats prevented from breeding, oestrus lasts about 10 days on average. However, mating makes the duration much shorter, ending it within a day of the breeding act or acts (repeated matings are usual). When there is no breeding, the eventual ovulations bring the heat period to an end. By contrast, mating causes prompt ovulations and these terminate the oestrus period. Among all the domestic animals, the cat is quite unique by ovulating on demand; that is, it has induced ovulation.

If the cat becomes pregnant and has her kittens, she sometimes comes back into heat when she has been suckling them for only a few weeks. However, the average time from birth to the next oestrus is 8 weeks. If the oestrous cat does not become pregnant, she usually comes back into heat in a range of 18–24 days. This roughly 3-weekly return to heat is likely to be repeated regularly throughout most of the year, although many cats stop their oestrus cycles for a while in early to mid-winter, namely the season when the days are shortening. Such anoestrous cats usually restart their oestrus cycles when the days provide them with significantly increasing quantities of brightness.

Indoor cats may not have the tendency to have a wintry rest from breeding since the species at large does not have defined breeding and non-breeding seasons. Tomcats are certainly on reproductive duty all their lives. Formerly all the oestrus cycles in the life of an unspayed cat were a great burden on both cat and custodian, but today a non-surgical veterinary treatment is available to suppress the cat's oestrus cycles for an extended period of many months and this is an aid to the animal's well-being.

Mating

The mechanics of cat coitus require an equal cooperation between the female and male participants. Their pre-coital phase involves a mutual lead-up to mating. In the presence of a male cat, the female exhibits her oestrous body language, but does not immediately accept the tom. The male continues to make tentative approaches to the queen in heat until he is able to mount.

The male cat's pointed penis is located below its anus and is directed backwards. In order for intromission to be effected, the tomcat must stand astride the female and arch its back tightly to position its penis towards the female's rear and its genital opening. The queen cat cooperates in effecting union by dropping on to its chest and raising its hindquarters to meet the down-turned hindquarters of the male. When there is genital apposition, the tomcat makes a single pelvic thrust to complete intromission.

As the penis is withdrawn, the reverse-facing barbed ridges around it evidently cause pain to the female cat, which immediately emits a scream and turns on the male to bite or scratch it about the head. A subdued relationship then follows for a short while until the copulatory process restarts. Several matings follow one another and different males may be involved in some of these. The female mating drive then dissipates and ovulation follows on the next day. Many breeding tomcats are left with bites on their necks and permanent notches on their ears.

In the female cat, coitus ends in aggression, not passively or euphorically. The sex drive returns only after a period of passivity. In due course the female's sex drive ends. Ovulation follows within 24 h; this allows the sperm to mature to have fertile capability, which needs numerous hours in the oviducts.

Pregnancy

The cat's length of pregnancy is usually 63 days. Its womb is a long V-shaped organ, often described as having two horns. In the most common pregnancy each horn holds two fetal kittens, but the number can obviously vary above or below this average of four. The pregnant cat's abdomen gets noticeably enlarged in the second month and her teats become more visible in the final week. In the days before birth the mother-to-be usually makes an effort to make or find a birthing place that provides seclusion. The cat retreats to this nest when the birth of the kittens is imminent. The birth site is used as a safe nest for the whole birth process.

Birth/parturition

All mammalian births have much in common. Cats comply with the generalities of fetal delivery. Initially there is the confinement. In this the cat takes its place for the approaching birth that is recognized in three stages, as follows.

1. The first stage birth is the period of painful surges in which the cat is absorbed with uterine contractions and the commencement of cervical distension.
2. In the second stage, there is full concentration on the muscular efforts to expel each kitten in turn, with the fetal membranes and severance of the kitten's umbilical cord.
3. Finally, there is attention to the cleaning-up of the second stage, including removal of fetal fluid adhering to each kitten, eating the fetal membranes and, in time, consuming the shed afterbirths. The cat then licks up the blood and other substances left over from all the births.

Stages 2 and 3 are repeated at the birth of each kitten making up the complete litter. A lapse of about half an hour may occur between each birth. Parturition is extended, depending on the number of kittens being born. On average, there is a delay of about half an hour between each birth, but a few queens may have as long as an hour or as short as 15 min between the delivery of each kitten. On extremely rare occasions, one uterine horn will empty itself one day, with the other uterine horn expelling its kittens on the next day. The delay between the deliveries of kittens allows the queen to attend to each neonate individually as it is born. Her attention to the freshly born kitten follows an attending procedure as follows:

- tearing the fetal membrane from the kitten;
- licking clean the kitten's muzzle and face;
- chewing off the kitten's umbilical cord;
- eating all the fetal membranes that have been expelled at that time;
- drying the kitten by grooming it overall, beginning at the head; and
- licking up all fluids associated with the delivery.

When the procedure is ended, the queen rests until the next kitten emerges. The attending procedure is then applied to the next kitten and similarly to all that follow. With a very large litter, the one or two at the end of parturition may not receive the same quality of attention and might even be ignored and left to die.

The cat's parturient third stage calls for much maternal commitment; in it she grooms the newborns very diligently. With her rasp-like tongue she scrapes off the thick fluid covering the kitten, paying particular attention to its head and its rear end (the feline tongue is like a flexible rasp). Grooming the face of the freshly expelled kitten helps clear its nostrils for ease of breathing. Licking its anus is believed to stimulate bowel motion in the kitten. This bowel motion by the kitten is necessary to expel the meconium that it has swallowed, forming the waste that has accumulated in the gut of the fetus during its gestation. This meconium is derived from swallowing amniotic fluid in the fetal sac and also includes intestinal secretions created during the later stage of gestation.

Prompt expulsion of meconium is vital for the health of the newborn kitten, by preventing a potentially fatal blockage of its intestines with that material. Some mothers consume the expelled meconium. When suckling starts, the mother positions herself on her side to allow the emerged kittens access to her mammary gland. When all the afterbirths are eaten, the cat's parturition is virtually ended. However, the mother usually transports her litter to a fresh, clean nest about 2 days after the parturition, as the final birth act.

Kitten survival

While no adequate survey has ever been made on the numbers and circumstances of newborn kittens that do not reach several days of age, we have different figures from various veterinary assessments that roughly estimate an average loss of 15%. Some breeds contribute more to that figure than others. For example, high levels of fetal mortality occur at parturition in the Manx breed, from a morbid, genetic factor. In breeds with very broad skulls or faces, the shape of the fetal head can impede the birth with a resultant difficulty in delivery and death of the fetus from prolonged, excessive pressure in the birth canal.

The surprising mortality rate in newborn kittens of all breeds includes approximately 7% that are stillborn and 8% that apparently fail in suckling. Evidently, cat litters have inherent problems of survival in which spacing and blind competition play parts. Kittens have their eyes and ears shut and their claws out for the first week or so. They have voices from birth and can cry out for maternal attention quite effectively from the start of life.

Successful suckling is a challenge for every kitten surviving the drama of birth. If any fetus has suffered trauma or heart strain during a protracted birth process, for example from having a malposture, the kitten may not have sufficient strength or energy to cope with the physical demands of getting attached to the mammary gland, finding an available nipple and sucking a full supply of the critically important first milk, while warding off challenges from adjacent, greedy siblings seeking to poach. Kittens compete seriously for sustenance. Any kitten that fails to master these tasks in suckling becomes cold, wanders away from the mammary region and dies in time. The mother may remove it from the litter as it progressively detaches from her and becomes cold. Survival of the fittest begins here, at the inaugural feeding session. Once the kittens are organized into a compact suckling unit they feed simultaneously (Fig. 6.1).

Fig. 6.1. Suckling unit.

Maternal role

The kitten's development is entirely dependent on its maternal relationship. From the moment of birth, the cleaning of the neonate by the mother is essential for the initial stimulation it supplies and the drying effect it has. Drying is necessary to prevent the development of hypothermia, which would lead to the death of the newborn kitten. The vigorous tactile stimulation created by the firm vibrissae on the mother's tongue arouses the neonate, which then becomes motivated by its instinctive neurology to seek the mammary gland of the cooperating mother. The positional movements by the mother begin when the last kitten has been born, preoccupation with labour has ceased and the kittens have been given their vital grooming. The litter is then organized into position by maternal adjustments.

The maternal motivation to suckle her neonates facilitates release of her milk, permitting the young to obtain the essential first feed. For the next 24 h the mother does not move from the birth site and she maintains intimate bodily contact with the litter to maintain its heat and to facilitate repeated suckling. As the kittens ingest, so they are required to excrete and the mother encourages this by licking their anal regions. She also consumes their waste, keeping the birth site clean. She continues with this practice until the kittens are about 3 weeks old. When they can eliminate their excretions effectively about that age, the mother demonstrates faecal coverage. Although the kitten is motivated by instinct to perform this feline ritual, some maternal tuition helps to establish the actual method.

Nursing maternal cats that are able to get outdoors hunt small prey such as rodents and birds to bring to the young kittens when they are a few weeks old. This introduces them to other food and it provides an opportunity to show such prey to them and how to catch it. This prey is brought to them in a live state and the mother releases it then recaptures it, showing the method of making a kill and allowing kittens to participate. Incidentally, this demonstration of catching and releasing prey by the mother is learned so well by most kittens that, in adult life, it becomes their own method of dealing with captured prey. Indoor cats do not perform this demonstration, but such cats, as a rule, do not have kittens, do not become mothers.

As the kittens become quite mobile, about the third week, they tend to stray from the nest and the mother then retrieves them by carrying them back by the scruff of the neck. Kittens respond to such carriage with the tonic immobility reflex that is detailed in Chapter 3. The mother provides a considerable amount of transportation to her kittens, since she usually moves them to fresh quarters several times during their first 2 weeks of life. Her guarding attention continues for some time, often beyond the weaning period.

The quality of maternal disposition in cats with kittens has a high modal value as outlined above, but it has variability dependent on the queen's temperament or her state of well-being. On rare occasions the mother abandons her kittens after their birth. Cannibalism by the mother on the newborns is known to occur in rare cases when it is her first experience of giving birth. If she is in poor health or undernourished she may be incapable of providing normal kitten care. Kitten mortality is high when the queen is unable to lactate adequately or to fulfil the maternal functions described.

The nutritional requirements of a pregnant or nursing cat involve increased quantity and quality of food. The volume needed can be three or four times her normal ration of moist food, perhaps up to 0.4 kg/day. Her diet should now include some foodstuffs beyond the usual contents of commercial shelf food. For example, there should be an additional supply of meat, liver, chicken hearts, egg yolk and some cottage cheese, all these in raw form. A supplement of vitamins and minerals requires to be provided two or three times weekly. Vitamin A is a special need. The vitamins and minerals needed for a nursing cat are listed in Table 6.1, with natural sources of these.

Table 6.1. Vitamins and minerals needed by a nursing cat, and their sources.

Nutrient	Source
Vitamin A	Raw liver, homogenized milk, uncooked egg yolk
Vitamin D	Fish oils, raw fish, butter
Vitamin E	Wheat germ oil
Thiamin	
Riboflavin	
Niacin	Raw liver
Pyridoxine	
Iron	
Copper	
Calcium	Milk, cottage cheese
Phosphorus	Raw meat

The relationship between the maternal diet and kitten health is very real. If the mother is inadequately nourished, the kittens will be malnourished or suffer clinical starvation. The various results are enduring on kitten health and include rickets, stunted development and shortened lifespan. Various other defects in well-being include poor learning of feline skills and behavioural disorders such as aggressive natures, or the opposite in extreme timidity. Kittens that spend long periods in futile nursing attempts tend to become antisocial towards siblings and this attitude often remains in the animal's adult character.

7 Giant Wild Cats

General

All 38 species of cats in the modern world are descended from eight lines of feline ancestors existing around 8–10 million years ago, a fact established through molecular genetics. This notable line of cats contains the genus *Panthera*, which, in turn, includes the three largest species of cats in the world today, namely the lion in Africa, the tiger in Asia and the jaguar in South America. Although there are other large cats in various locations, including the puma in North America, the cheetah in Africa and the leopard across Africa and Asia, the three former species are the giants in the feline family, with weights exceeding 100 kg. The tiger can weigh up to 227 kg, the lion can be up to 190 kg and the jaguar can reach over 100 kg. Vaillant (2010) comments that Siberian (Amur) tigers can weigh from 227 to 317 kg. To put this in perspective, a tiger of this scale would be about 50 times the size of an ordinary housecat.

In addition to sheer size, these huge cats share much in common, as well as having differences. Their coats, for example, exhibit distinguishable dissimilarities (Kitchener, 1991). The tiger is the only striped cat; the lion is plain coloured; the jaguar is heavily spotted with large rosettes. Lions live in groups termed prides, while tigers and jaguars live a life of solitude. Retractable claws are a characteristic feature of both tigers and jaguars; lions, however, do not possess this ability. When hunting, lions and tigers immobilize their prey by biting the nape or throat; jaguars, however, kill their prey by piercing through the skull. Lions hunt together in groups, while tigers and jaguars search for prey alone. All three feline species can swim and exhibit proficient fishing skills. When catching mammalian prey on land, the lion consumes its kill in the location where it was brought down. Unlike the lion, the tiger and the jaguar usually drag their catches into dense cover before settling to eat them. All three felines hunt through the same method; stalking prey until reaching about 20 m of its position and then charging at the animal. Success rates per attempt are usually below 40%. If prey is large enough to flee at speed, the predator has no choice but to give up the chase. These large carnivores can maintain top speed only for 100 m or less before exhaustion prevails.

Lions, tigers and jaguars will kill and eat any animal they encounter, ranging from very minute animals to large ungulates. Commonly, they will prey upon animals much larger than themselves. All three felines breed in the same manner and may suckle their cubs for about 6 months. Lion cubs exhibit their own technique of feeding by suckling any female in the pride that is lactating. Relatively long faces, large wide noses, almond-shaped eyes and small rounded ears characterize the facial features of all three cats. As would be expected of giant felines, all species demonstrate great muscular power and ability. Both the lion and the tiger have a similar physique, with well-developed forequarters and hindquarters of a finer build. All the wild cats of the world, including these three giants, have been extensively reviewed by Sunquist and Sunquist (2002) in their encyclopedic book *Wild Cats of the World*, which is used in the present text as a basic resource, together with Wikipedia, 2010.

Lion

The lion (*Panthera leo*) is commonly regarded as one of the most dangerous animals in the world (Guggisberg, 1975), but evidence suggests otherwise. With the phenomenal benefit of early sensitive human handling (to be discussed in Chapter 11 as ESH), even lions can become affiliated with humans (Bourke and Rendall, 2009). Ancient cave drawings depict the lion coexisting with mankind. The Bible makes it plain that lions existed in Palestine 2000 years ago. During the Roman

Empire, lions were imported from North Africa for public viewing and to serve as pets. If they are socialized while young they can become quite tame (Fig. 7.1).

They were housed in European menageries for public interest in The Middle Ages. At this time, lions lived in various European regions. Until two centuries ago, lions still resided in many parts of the Middle East, including Syria, Iraq and Iran. As a result of deforestation and hunting for sport, their numbers were eliminated. Today, the lion exists in significant populations only in sub-Saharan Africa. A small population of Asian lions (*Panthera leo persica*), however, has been re-established in the Gir Forest region of north-west India, following efforts to continue the existence of the Asian subspecies of this dramatic cat. The need for conservation of the great cats has been appreciated for some time (O'Brien *et al.*, 1987; Nowell and Jackson, 1996; Sunquist *et al.*, 1999) and most recently by Weston (2011).

The lion illustrates the status that cats can hold in their particular habitats. This feline occupies the extensive grasslands of Africa that extend south of the Sahara, down East Africa, then southwards and across the lower part of the continent, notably the Serengeti and the Kruger (Kingdon, 1977; Smithers, 1983; Stuart and Wilson, 1988; Mills and Biggs, 1993). Lions are regarded as principal members of the family Felidae. They are a protected species,

(a)

(b)

Fig. 7.1. (a) Author's daughter, Sheila, with lion. (b) Author's granddaughter, Claire, in lions' den.

indicating their vulnerability to the ways of mankind. This vulnerability applies to all cats, in spite of the caution and defensiveness that characterize their typical feline disposition.

Characteristics

Males have a height of about 120 cm at the withers and weigh about 225 kg, while females are smaller, about half of a male's weight. The coat is in various sandy-coloured shades, including yellowish brown, reddish brown, light buff, dark brown and grey.

The underparts of the body are usually of a paler colour. The end of the tail is characterized by a black tuft, while a large black spot is notable at the back of each ear. A most outstanding feature is the dense and ragged mane carried around the neck of an adult male, following its fourth year of life. In the younger adult, the mane is a yellowish brown, but darkens and becomes a thick mat with age. The mane gives an adult male its distinguishing appearance and thus its identity. The head is proportionally long and wide, with small rounded ears and long canine teeth. Females also have large heads (Fig. 7.2).

Lions inhabit a variety of environments including desert, grass plains, wooded grasslands, forests and scrub lands. Although they live terrestrial lives, lions are known to climb trees occasionally, resting along a branch during the heat of the day. Lion cubs have also been observed in trees, engaging in play. While indolent in the daytime, lions become more active at night. This is not, however, a definite rule – there is seasonal variability in their travelling and hunting times, and they avoid intense activity during the hottest season. In the Serengeti, a notable domain of lions, the peaks of activity are dawn and evening. Dark nights usually provide the best hunting results. Lionesses may spend as much as 20 h resting, while travelling and hunting for the remaining 4 h of a day. Travelling distances may only

Fig. 7.2. Sexual dimorphism of the lion's head; in the old male (top), note the greater mane.

amount to about 10 km where prey is adequate in the district.

An outstanding feature in lion behaviour is the formation of social groups, termed prides (Fig. 7.3). The only other feline to form social groups is the domestic cat, particularly when it goes feral and gathers into colonies (Passanisi *et al.*, 1991). Lion prides in the Serengeti, for example, are usually composed of a group of closely related females and their offspring, as well as a few subadult and adult males. The pride may total any number from seven to 35 individuals. Small prides may have only one or two males and five females. Each pride has its own territory, ranging from about 60 to 180 km^2. Kills are shared by pride members (Estes, 1992).

Predation

The principal prey animals of African lions are ungulates, particularly wildebeest. Other prey includes larger mammals such as giraffe and Cape buffalo, as well as smaller animals including hares, birds and reptiles. In a pride, most of the hunting is performed by the lionesses, while pride males get priority at feeding. Lions usually ambush their prey by hiding in long grass, often in close proximity to a watering hole. Prior to the final rush, they will creep quietly to within 25 or 30 m of the animal. Lionesses hunt cooperatively, with differing pride members assuming certain vantage points. At times, females from other prides will assist in the hunt (Elliott *et al.*, 1977; Van Orsdol, 1984; Stander, 1992). When the prey is knocked down, the lion may grasp the animal by the muzzle before biting the neck.

With several lions involved in a kill, consumption follows fiercely, quickly reducing a large ungulate carcass to a skeleton. The soft viscera and fleshiest meat are eaten first. With the possible intrusion of scavengers such as hyenas and vultures, a carcass must be stripped with haste. Small prey items are usually carried to a safer location before they are eaten. Success rates in hunting are often about 30% and invariably less than 50%. This success rate is higher than is usually achieved by other wild cats. Hunting in groups appears to facilitate larger catches. A carcass is so quickly devoured by the pride, however, that no single lion receives the full benefit of a kill. Sometimes, two lionesses will hunt together and share their kill, but without the help of the pride, their ability to hunt larger animals is reduced (Mills and Biggs, 1993).

Association

Separate coalitions of adult males may number from two to six, and will usually hunt together. These coalitions may exist separately for long periods. At times they will join a pride, but are met with attempts of rejection by females, as invading males have a tendency to kill young cubs. An adult male established in a pride may remain with it for several months or years. The lion's social system is not always fixed, but is often fluid, with members moving to its perimeter or leaving to join another pride. Males in the pride come and go, while subadult males remain in the pride for lengthy periods, leaving when they mature to join male coalitions that may endure for about 2 years. The feline

Fig. 7.3. Pride of lionesses.

dispersal phenomenon operates to cause this departure of males, while young females normally remain in the natal pride. The density of lion numbers within their home territories varies according to the available biomass. In some areas there may be a couple of lions per 100 km^2, while in prey-rich lands there may be 50 lions per 100 km^2.

Territorialism

The feline practice of marking home territory is carried out assiduously by male and female lions with scent-marking. This is performed by brushing their hindquarters against a tree or bush and spraying urine on to surfaces. Another marking system is scratching the ground with the hindfeet and then spraying urine into these marks. Certain landmarks in their territory such as tree trunks and rocks are sprayed repeatedly. With one lion actively scraping the ground or spraying an object, others in the pride often follow suit to make their marks. In claiming territory, lions make use of their ability to roar loudly, while standing upright. The sound of roaring can carry for great distances and is very effective in advertising the lion's particular location. Most roaring is done at night when the animal is active. The protection of territory by these various means reduces fights with travelling males that may wander into an established territory. When a lion starts to roar, it encourages other members of the pride to participate, creating a harmonious forum.

A lion pride can be quite large and may overlap the territory of another pride. When this occurs, males patrol the perimeter of their territory, defending it against intruding lions, and females guard such parts of their territory that contain watering holes and den sites. Pride members cooperate in maintaining their home range exclusivity. Some lions, including females, do not join a pride and exist in a nomadic fashion without any specified territory, existing quite successfully. Lone females, however, do not have much success in breeding or in raising their young if they do breed.

Breeding

When a lioness goes into heat she performs the standard feline oestrus display; rolling on the ground, twisting and turning, calling and taking brief walks. The closest pride male attending the female has priority at breeding and as a result, pride males seldom fight one another for access to a breeding female. This male, likely awaiting the development of oestrus having detected proestrus in the female, now displays more attentiveness to the lioness and wards off other males. The pair may exchange social licks. At the start of her oestrus, the lioness rejects the male, growling and batting with a foreleg. The male will not make any further, serious attempt at breeding until the female shows her receptivity several hours or even a day later (Packer and Pusey, 1982).

Copulation begins with the female performing lordosis – sinking her back, turning her tail to one side and raising her hindquarters with her belly on the ground. The male sets himself astride the female, arching his back to mate with her. Intromission is brief, lasting only seconds. As the male withdraws, the female emits a cry and turns to bat or bite the male. After a period of rest, copulation is performed again, and this repetition of mating may continue for numerous copulations over a couple of days. The original male eventually ceases to remain close to the oestrous female and, in his absence, other males begin to participate in the ongoing breeding operation. The female may remain in oestrus for about a week, mating frequently during this time, often with several males. Not all males present in the coalition become involved in breeding, however; particularly younger and older males do not become involved. It is common for several lionesses in the pride to have synchronized oestrus cycles and to breed more or less at the same time.

The gestation period is about 3.5 months. Litters are born in a secluded den of the mother's selection and usually vary in number from one to three individuals. The newborn cubs appear deeply slumberous for about 3 days before becoming active. Remarkably, some cubs open their eyes soon after birth, while others cannot see until they are about 10 days of age. Cubs at this stage are helpless and vulnerable, and the mother will be as protective as circumstances allow. Nevertheless, many lion cubs become prey to scavenging animals including hyenas, leopards and even other lions. Mortality rates are high among cubs as a result of these predators and early starvation, if the mother does not have an adequate supply of milk. A factor modifying starvation is the practice of nursing cubs on non-parental lactating females. The phenomenon of synchronized breeding results in several females giving birth simultaneously, therefore nursing at the same time. The young cubs in the pride benefit from the availability of several lactating females willing to suckle them. In the business of communal suckling, small cubs are at a disadvantage due to the strong competition of numerous cubs fighting to obtain a teat.

Cub development

The maternal lioness shelters her cubs in the den for the first 6 weeks. Remaining here continually for the first few days, she is soon forced to leave the litter to obtain food for her own need. Leaving the cubs unprotected is a hazardous time for the newborns. Once they are able to walk, she may try to relocate them to a new den, but infant cubs are unable to travel far. By 6 weeks of age the cubs are now physically able to accompany their mother when she hunts. They learn to eat solid food but continue to depend primarily on milk until they are about 7 months old. When they are still nursing, small cubs have difficulty obtaining enough to eat from carcasses because of the furious competition with others, including adults. Lactating mothers collaborate with the general defence of cubs, but non-lactating females do not. Survival can be difficult for cubs between 7 months and 1 year of age, and it has been reported that a mortality rate of 28% from starvation can occur in this age group (Schaller, 1972). When prey is scarce, lions travel much further than their normal territorial limits in search of food. While adult lions can walk long distances, cubs are often unable to keep up with the pride and are left behind on their own to die. Maturing cubs stalk prey and observe adults during hunting, but do not possess the strength or weight to hunt down large prey until they are about 1.5 years old. Proficient hunting is basically dependent on instinct, learning and physical development. The paternal lion participates in the guarding of his cubs (Fig. 7.4).

The demanding existence for lions as they mature results in some mortality. Schaller (1972) estimated that about two-thirds of all lion cubs in the Serengeti die each year from a variety of such non-disease causes as starvation, predation, infanticide and abandonment.

Pride dynamics

Occasionally, a coalition of younger, stronger males will take over a pride. Male lions occupying the pride are ousted in the process, with some enduring injuries. The displaced males often become nomadic, travelling across various territories for the remainder of their lives. The new males are a serious threat to the young cubs, preying upon them. Fathers of cubs defend them. In an attempt to defend their cubs, females are often killed. As a result of this risk, lionesses with young may also abandon their pride to travel on their own. Death of their young causes the mothers to come into oestrus in a matter of 2 to 3 weeks. The new males remain within the pride for long periods, breeding with female members and accepting their own litters without infliction. Females typically stay in the group for life, although some may choose to leave and return later. Lionesses with successive

Fig. 7.4. Male lion guarding cubs; note expression.

daughters may stay in one territory for generations. Female cubs will remain with their mother's pride as they become subadults, but male cubs will leave when they reach sexual maturity. This stage cannot be reached, however, until about 4 years of age. In the wild, female lions have a lifespan of about 18 years and they may breed until they are 15 years old – they are, however, less fecund in old age. Male lions seem to have a shorter natural lifespan than females. In both sexes, longer lives are often attained in captivity. Efforts are now being made to conserve lion groups in parts of Africa for tourist viewing (Sunquist *et al.*, 1999; Thornton, 2011).

Asiatic Lion

In the Gir Forest of India the Asiatic lion has now been re-established in a large population through the Indian government's attempts at conservation. The Asiatic subspecies is very similar to the African lion in appearance. One notable difference is the prominent belly fold in the Asiatic; this is a thick flap of skin running the length of the ventral region from the chest to the prepubic area. In addition, the Asiatic lion has a thicker coat and males have thinner manes. The two species are similar in all other respects, including their principal choice of prey (ungulates). In order to provide the lions in the Gir Forest with appropriate prey, the State government removed farmers and their livestock from the area to allow space for wild ungulates to graze, without competition from cattle and goats. As a result, there was an increase in the populations of indigenous species, particularly chital, sambar and nilgai. The latter is a large antelope. The lions preyed on these animals and the leonine population flourished (Sunquist and Sunquist, 2002).

Inevitably, the predator population increased to the point that lion numbers exceeded their forest territory and ultimately invaded human space. Human attacks became frequent and deaths numbered 23 in 1992. Farmers within 5 km of the Gir Reserve lost increasing numbers of their livestock to these lions. The conservation planners for these Asiatic lions clearly ran into a major difficulty as a result of their own success. Translocation of captured lions proved to be a failure, since many individuals returned to their original locations. The only solution was shipping excess lions to zoos, where they were welcome as Asiatic specimens. A management scheme was also introduced to compensate farmers for their livestock loss. Local people came to accept the merit of the Asiatic lion conservation scheme when it became controlled. This scheme has set an example for plans in other locations to preserve wild cat populations. Protection of a limited number of individual carnivores can be an initial success, but can lead to the overpopulation of predatory animals – a problem too difficult to accommodate in any given space.

Tiger (Siberian and Bengal)

General

The tiger (*Panthera tigris*) inhabits part of the Asian landmass as two species, the Bengal and the Siberian. Tigers are the only cats with stripes, and are biologically closer to the domestic cat than are lions, sharing retractable claws and superb night vision. Their incisor teeth are in straight rows and their whiskers project prominently, as in most housecats. The tiger's striped coat slightly resembles the striped tabby.

The Siberian tiger (*P. tigris altaica*) can be found in eastern Siberia and northern China. It is the largest cat in the world – an individual has been recorded weighing roughly 382 kg (Vaillant, 2010). Siberian tigers often range extensively in their hunts, travelling hundreds of miles in search of large prey such as deer, bear and elk. Smaller catches include hares, rabbits and fish. The male lives a solitary life, while females live in small family units. The territory of a single male can extend over 7500 km^2. The Siberian is tolerant of other tigers, male or female, crossing through its home range, as it is not particularly aggressive outside of hunting. If portions of its kill are left over following the first feeding, it typically scratches soil and litter over the remains to be uncovered later and eaten. The Siberian tiger has become a rare sighting and is listed as an endangered subspecies. Some individuals are preserved in zoos as wonders of nature. With their relatively low aggressiveness, they readily adapt to zoo life (Fig. 7.5).

The Bengal tiger (*P. tigris tigris*) has its habitat in central and northern India, Bangladesh, Burma and Nepal. A century ago, it flourished with its greatest population of about 50,000. Less than 10% of that number constitutes its current population, but a conservation programme for the animal is now successfully in effect. The Bengal tiger corresponds to the African lion in size and predation, but its habits are in contrast – characterized as cat-like.

Fig. 7.5. Zoo tiger resting comfortably.

Its lifestyle is solitary and nocturnal with a distinct springtime breeding season. The Bengal tiger and the housecat share a number of similar features; their vocalizations are of a comparable variety, both have small litters and enjoy using an array of resting places within their territories. The males of both feline families spray urine to mark their home areas as indicators for others that may intrude. Although lions and tigers roar, the Bengal's roar is uniquely loud.

Bengal tigers prey upon water buffalo and wild oxen chiefly, but also hunt wild pigs, deer and monkeys. Like the Siberian, the Bengal exhibits the habit of covering unfinished meat and returning later to finish it, even if it has become putrefied in the interval. This latter manner of feeding is in marked contrast with other cats and the domestic cat in particular. The housecat is fastidious in its eating style, selective in its choice of food and often hard to please, particularly with titbits.

An obvious difference between the Bengal and the Siberian tiger is in their thermoregulation. The Siberian must deal with ambient winter temperatures of about –40°C, while the Bengal experiences near tropical heat all year. The Siberian develops a heavy winter coat and pads of subcutaneous fat to insulate its body core. The Bengal seeks shade in the forested areas of its natural habitat to avoid direct solar radiation in times of excessive heat. In addition, Bengals often bathe in pools during very warm periods. This habit is rarely observed in the domestic cat – at most, a housecat will allow its paw into running water.

Siberian Tiger

Siberian tigers were once distributed extensively across the far east of Russia, northern and eastern China, Korea, the Tibetan Plateau and Manchuria. They were variously named the Korean, Manchurian, Ussuri, Woolly or Amur tiger, and their population never exceeded 1000. Presently, they are classed as Siberian or Amur tigers, and their scattered populations in north-eastern Asia have been drastically decimated by hunting. Siberian tigers now exist in just a few scattered locations on this land mass. Vaillant (2010) has provided a full account of their circumstances today in the lands of Amur and its surrounding areas. The prognosis is bleak – a great animal has been brought to the brink of extinction as a result of mankind's activities, namely hunting, poaching and the removal of habitat.

Vaillant (2010) notes that the estimated number of tigers throughout Asia a century ago was 75,000; recently, their numbers totalled just 3200. He points out that 95% of the tiger population was killed for sport, beauty, medicine, money, territory and revenge – a wipe-out that occurred in one century, following a thriving existence and ecological harmony with its environment for millions of years. The majority of the tiger's decimation took place between 1880 and 1960, the principal years of what was called 'big game hunting'. In spite of this, the early 1940s saw a wide distribution of these tigers throughout Asia. They could be found almost anywhere on the continent, from north by Vladivostok, to south by Bali, east by Hong Kong, and west by Iran. Only within the past few years have four or five subspecies of tiger become completely extinct in their regions, namely Bali, Java, south China and the Caspian region. In addition, there have been no tiger sightings in the Koreas since 1991. Manchuria, the presumed birthplace of the tiger, is also without this magnificent animal.

Characteristics

The Siberian tiger is the one suited to living in the arctic conditions of the Russian far east. It roams such great distances during its lifetime, particularly in the winter season, that it was once considered a pure nomad. On the eastern edge of its true domain, namely the far east of Russia, it wanders on foggy shorelands and beaches. A very capable swimmer, this tiger is able to hunt ducks at sea and catch salmon and seals. On land, its prey ranges in all sizes from small rodents to adult bears. A full-grown bear should be too formidable an opponent for a tiger, but many reports indicate tigers having killed black and brown bears weighing about 450 kg. In addition, bears are known to flee from the sight of a tiger.

The Siberian tiger in winter carries a heavy coat that can reach the ground. It has a large head, being foreshortened and extremely broad while surrounded by a thick neck (Fig. 7.6a and b). Its canine teeth are about 8 cm long. The paws are huge and enclosed in fur; the forepaws being larger than the hindpaws. When the forepaws are extended, the claws are extruded like knives. Its head and body length is sometimes more than 2 m and it is over 1 m in height at the shoulders. In former times, some Siberian tigers were estimated to weigh about 364 kg. Vaillant (2010) quotes reports in reputable publications of Siberian tigers in the past weighing up to 408 kg. Such huge tigers were prime game for hunters, and these large felines now appear to have been eliminated as a result. The largest Siberian tigers now weigh about 220 kg.

With their great size, tigers dominate their ranges. In some regions they coexist with leopards and wild dogs, but these differing species tend to avoid each other. Leopards, in particular, refrain from such encounters. In turn, Siberian tigers avoid larger mammals such as adult rhinoceroses and elephants. When a tiger tackles a large buffalo or a gaur (bison), the outcome can be success or failure; both of these large ungulates have been known to kill in the course of a tiger attack. The Siberian tiger in the north, however, frequently kills moose weighing over 450 kg, such is its power.

(a)

(b)

Fig. 7.6. (a) Features of tiger heads; note large nose (top) and camouflage effect (bottom). (b) Watchful Siberian tiger, world's largest cat.

Predation

The Siberian tiger stalks carefully (Fig. 7.7a and b). It has two methods of killing. For small prey, a bite on the back of the neck severs the spinal cord. It is believed that the tiger can feel its tusks into an intervertebral space to break the cord in the manner of pithing. With large prey, the tiger bites the animal's throat, crushing the trachea and causing suffocation. In this process, the tiger retains its grip on the animal's neck after the prey has stopped struggling. If the animal manages to escape the initial attack, the tiger will not pursue it. Its hunting method is simple: it pounces quickly and forces the animal down. While the Siberian tiger can walk for days when prey is scarce in winter, it cannot run for a long distance without experiencing early fatigue (Vaillant, 2010).

As a predator, the Siberian tiger is not as ravenous as the wolf; tigers hunt for necessity, while wolves kill excessive numbers of animals ruthlessly. The tiger maintains a healthy environment by controlling imbalances in populations. It could also be regarded as an agent of evolution, by eliminating the poorer specimens of any given species – those that are easiest to catch. The theory could be advanced to suggest that the sabre-toothed tiger's predation created the horse, an ungulate of modest proportions originally, that became highly developed for power and flight. Studies have shown that feline predators do not diminish the total biomass of a given environment. The greatest ungulate biomass in southern Asia, for example, is located in typical tiger territory where forest and

Fig. 7.7. (a) Stalking style of tiger; note placement of right forepaw for stealth. (b) Prowling tiger using bush as a screen.

grassland form a mosaic (Sunquist and Sunquist, 2002). Amidst the predators here, the biomass is stable.

Where all sizes of prey are abundant, the Siberian tiger chooses to target the larger animals. Such kills can provide a substantial amount of meat, sustaining the tiger for a week and thus eliminating the need to hunt again during that time. In hot weather the carcass decomposes quickly, but such advanced states of decay do not deter the tiger's appetite, as it will continue to consume a carcass until it is finished. The remains are found as bones scattered over a wide area. Large kills can be carried away over considerable distances. Sunquist and Sunquist (2002) provide accounts of one Siberian tiger dragging the carcass of a bull weighing 770 kg for a distance of 500 m; another tiger was observed dragging away an adult horse; yet another is described as carrying a cow for 2 miles. Typically, the tiger moves its prey into a secluded, shaded location before settling down to eat. When the tiger ingests its catch, it typically starts at the animal's rump. It then tears open the abdominal cavity, removing the stomach and gut before devouring the remaining viscera. At this time, it may relocate the carcass to a fresh site. Eventually, all of the soft tissue on the prey is eaten. Sometimes, Siberian tigers will share a kill with another.

It is estimated that a full grown tiger hunts a total of 3000–4000 kg of prey per year, dependent on whether the feline is a single female, a mother with cubs or a large male. Schaller (1972) observed in Chitwan that the biomass of prey species increased in one year, while the resident tiger population in the same area remained stable. Where prey numbers are limited, however, a given species can suffer from excessive predation. Karanth and Sunquist (1992), for example, found that solitary eaters, such species as sambar and barking-deer, are more sensitive to predation and may suffer a drop in population from tiger predation. In the same observations predation was found to be heavy on chital, gaur and wild pigs, but their gross populations were not affected by tigers. The conclusion from similar findings elsewhere notes that despite short-term fluctuations in prey numbers, tiger predation does not affect their population size in the course of an extended period of time.

Siberian tigers hunt and stalk alone although an encounter with another tiger may bring them together to share a kill. Individuals associate at breeding time, but only for a short period. The only lasting relationship is the link between a mother tiger and her young. This family group remains close together until the cubs are about 1.5 years old. Tigers, within their society, communicate by occasional encounters and vocal signals. They determine the whereabouts of others through scent-marking deposits. Regular travellers, they sometimes cover distances of hundreds of kilometres in one excursion, particularly when prey is scarce. The Siberian tiger is known to travel extensively in winter (Vaillant, 2010). The tenure of their territory is quite flexible and dependent on the nature of the environment (Seidensticker and McDougal, 1993). A tigress with cubs, however, is very defensive of her small territory, which usually spans between 10 and 20 km^2. She usually remains for life in a territory that contains sufficient prey. Female territories do not usually overlap.

Territorialism

For both Siberians and Bengals, male territories are held exclusively and are huge in comparison with those of females. Ranging nearly 200 km^2, these male territories can include up to seven female territories (Sunquist, 1981). Females seek areas with food, water and shelter for their young. Reproduction has the highest priority in tiger motivation, with food a close second. Male tigers regularly tour their ranges, marking them with scent and avoiding the territory of any other male. Although they have their own limits, tigers are quite aware of the existence of their neighbours.

Sunquist and Sunquist (2002) summarize marking effects very comprehensively by noting that, although tigers are asocial, their lives are largely governed and regulated by a complicated mixture of chemical and physical signals. In addition to marks, tigers communicate vocally and the variety of their sounds is extensive. They utilize some sounds for long-range expressions and others for use at close range, even in face to face positions. In the latter situation, grunts and puffing are communicated as friendly greetings, often used by the maternal tigress towards her young. To express their presence, tigers also growl, snarl, grunt, moan, hiss and miaow. On rare occasions they emit an unusual coughing snarl or pooking sound. Their roar is atypical – a drawn out, rolling, deep howling expression. This roar variation can carry far, and can be detected by the human ear over a distance of 3 km. The tiger's type of roar is used in a variety of situations and may not have specific meaning beyond an expression of assertion, or to indicate the animal's gender.

Tigers use a range of methods to mark out their territories; they create deep claw marks in tree trunks with their forepaws; they scratch the ground with their hindpaws; they leave deposits of faeces on prominent sites in their home areas. Chiefly, tigers spray a mixture of urine and musk on to rocks, trees and bushes – surfaces that will retain this form of signage for a length of time. Following the death of a tiger, its territorial marks remain effective for several weeks. Neighbouring or wandering males in search of a new territory will refrain from the deceased tiger's territory for several months.

Bengal Tiger

Sunquist and Sunquist (2002) state that contained in territorial marks is information regarding individual identity, sex and reproductive condition; scent marks serve to bring animals together for mating purposes. A tigress's rate of scent-marking increases to twice the usual incidence just prior to her oestrus, and while she has a mate her marking appears to end. Despite their larger territories, male tigers are diligent in maintaining their marks by revisiting every portion of their space on a regular schedule, depositing fresh marks every 2 or 3 weeks. This procedure also keeps the male informed of the reproductive status of the females within his territory. Females also mark their territories on a regular basis, covering their entire territory every few days. An increased rate of marking announces that her oestrus is imminent. In response, the resident male is soon present. On meeting, the female bites the male gently before rubbing against him. Familiarity is an aid to receptiveness between the pair, who remain together for only a few days of mating.

Tiger territory is constantly probed by young dispersed males wandering in search of a place where they can survive alone. Cubs spend 1 or 2 years with their mother, learning how to hunt and kill for themselves. When the dispersal phase sets in, the male cubs are suddenly without a home or territory. Although they are well grown at this time, they are still immature animals, wandering without direction and seeking a habitat free of competition, containing enough prey to support them. In the course of this difficult search young males encounter numerous hazards, regularly involving conflict with other males. Fights ensue, from which many tigers die. A young male that eventually gains its own territory is an extremely capable animal, having coped with major tests of endurance (survival of the fittest). In this process of self-establishment, most young male tigers have an unfortunate end.

Dispersal

At dispersal, a female cub avoids this process by remaining in the natal area for life and acquiring a comparatively smaller territory close to the maternal range. This territory will fall within the greater span of a male territory, which, in turn, will provide considerable protection from any intruding alien tiger. While the lion's method is pride formation, the female tiger has an equivalent system of clannishness. Females occupying territories within a broad area are likely to be related and without competition. They have knowledge of their communal relationship and repulse any unknown female entering the area. Both systems, of female prides and female clans, are effective strategies for optimal reproduction within each of these species.

Man-eating

The tiger's reputation for hostility towards humans is fuelled by the few Bengal tigers that engage in man-eating. Vaillant (2010) quotes a scholar who estimated that tigers have killed a million Asian citizens over the past 400 years. These occurrences were spread across many countries; the majority, however, were in India, as a result of the Bengal tiger. Certain individual man-eating tigers killed scores of humans in India before they were tracked down and shot. In other lands such as China, Korea and Manchuria, man-eaters were also a scourge. Their vicious reputation was taken as a justification for hunting the species as fair game. A Tiger Hunters Guild, for example, was formed and operated in north-east Asia, killing hundreds of tigers, chiefly for self-enhancement. In 1911 King George V and his retinue went on a tiger hunt in Nepal, exterminating 39 tigers in 10 days (Sunquist and Sunquist, 2002). One Indian Maharaja claimed to have killed 1150 tigers during his hunting career. Engaging in tiger hunting trips appeared, for a while, as high fashion in Western society.

The view held today is that tigers do not regard humans as prey unless the animal has experienced trauma as a result of human hands. Such trauma is often considered to be wounds from firearms, evident in old injuries found in many man-eating tigers. In the dramatic true story by Vaillant (2010) of a hunt for a man-eating Siberian tiger, the animal at the centre of the story, when eventually killed, was found with several old wounds. It is suggested by many tiger cognoscenti that if a tiger becomes physically unable to hunt properly as a result of injury from fighting or being hunted, it may resort to human attack. Tiger hunting for sport is now banned in most Asian countries, including Russia; despite protection, illegal poaching has been difficult to control. High prices are paid in China for certain tiger parts that are believed to have medicinal value. This belief is strongly established in Chinese culture and will continue to have a negative effect on tiger welfare, unless the tiger's rate of reproduction can equal its mortality rate from poaching and natural causes. Plans are in effect for conservation of the Amur tiger (Sunquist, 1997; Vaillant, 2010).

Breeding

Knowledge of tiger behaviour primarily concerns the Bengal tiger, well documented by Sunquist and Sunquist (2002). The Bengal tiger's breeding process has been well detailed, indicating essential similarities in reproductive behaviour within all felines. Mating tigers follow a ritual; although she is already in oestrus, the female resists the initial approaches of the male by spitting, snarling and striking him with her forepaws in mock aggression. The male's response to this resistance is passive avoidance. Like their Siberian cousins, the oestrous Bengal female continues to display all of the behavioural signs of oestrus; rolling on the ground, calling and presenting her hindquarters in the male's direction. In turn, the male makes occasional approaches to the female. With a familiar mate, the tigress eventually ceases her aversive displays and accepts the male, biting him gently and taking a stationary, squatting position; this posture of lordosis is adopted with the tail raised to one side.

The male tiger mounts the receptive female, crouching over her without applying any pressure. Arching his back, he directs his genital area to meet her raised hind position. Genital apposition permits intromission by the male, followed by a short series of pelvic thrusts as the mode of copulation. This motion may last just 15 to 30 s, with the male retaining a bite on the neck of the female. As the male begins to dismount, the female emits a growl and turns on him quickly, in the attempt to strike or bite his head. The two withdraw from each other for a short distance and remain recumbent. Following an interval of 5 min the pair mates again in the same fashion. Copulations continue frequently, up to 50 times a day for several days, until the male decides to discontinue and leave the female. Another male may then advance and participate in the breeding process. In order to induce ovulation in the female, numerous copulations are required.

Cub development

The tiger's gestation lasts about 3.5 months. As the birth is approaching, the female seeks out a den, often choosing a rocky opening or a dense patch of grass. Commonly, litters consist of two to three cubs, but may also range from one to seven. During their first few days of life, the cubs nurse their mother and sleep without any significant activity. The mother remains with her young continually for the initial days following the birth, at which point she departs on brief, daily searches for food and water. The cubs open their eyes at roughly 10 days of age. When the cubs are 1 month old, the mother diminishes her suckling time to about half of what she provided initially. By this time, the cubs have increased their weight significantly and are actively exploring beyond the den. They are finally weaned from their mother when they reach about 3 months old, having been introduced to fresh kills at about 6 weeks of age. Some cubs attempt to continue suckling until they are 5 or 6 months old, but most young tigers become completely carnivorous at an early age.

The maternal female is extremely secretive and protective of her young, moving them to a fresh location if a threat is detected. She is attentive when leading the cubs to accompany her on hunting trips, ensuring they are able to maintain their energy on these excursions. They follow her movements and observe her in action but do not participate in the hunt at this stage. At 6 months of age the cubs are not yet able to hunt for themselves, and rely on the catches provided by their mother. The young are quite efficient at sharing the mother's kill, and will continually depend on her until they are 1 year old; they are, however, active participants in their mother's hunting by this age, and even continue to hunt with her until they are a little over 1 year old. By this time, the male cubs have grown sufficiently to hunt and kill on their own. An 18-month-old male tiger cub can weigh about 150 kg, while a female cub at an equivalent age can weigh roughly 110 kg.

Dispersal

Young tigers become independent of their mothers between 1.5 and 2 years of age. After leaving her they will, however, continue to live within her territory for a couple of months. By this time their mother may be caring for a new litter, and this onset appears to be a key in the dispersal phenomenon. The juveniles do not necessarily leave their mother's domain if she has no subsequent litter; however, dispersal is the norm at about 2 years of age. Females have greater success than males in locating new territories, and as discussed, such territories are often in the general neighbourhood of the natal area. Males travel further than females in their search, often covering hundreds of kilometres. In the course of their quest they often wander into established territory in the search of food, and eventually encounter older males. The resulting fights usually do not end in favour of the young animal and it may be injured or killed in the inevitable battle. Occasionally, a younger tiger will confront an aged tiger that still holds a territory, prevailing over the older animal in the encounter and thus succeeding in gaining the territory of the former occupant. The latter is likely to abandon his claim in the dispute and become a nomadic tiger for the remainder of life.

Longevity

Female tigers (both Siberian and Bengal) can breed throughout most of their lives and many live up to 20 years of age. Males, in their natural environments, have a reduced lifespan as a result of injuries most likely caused by fighting. These injuries cause failure to hunt successfully, leading to starvation and death. In captivity, there are thousands of healthy tigers living long lives. Eventually, the numbers of tigers inhabiting zoos will surpass the numbers residing in the wild; this is already a fact with the Siberian tiger (Vaillant, 2010).

The alternative to zoo accommodation is the establishment of tiger reserves, but such reserves need to have fixed security (Sunquist *et al.*, 1989). In the Sundarbans of India and Bangladesh, for example, tigers have some protection. This is an area covering 14,000 km^2 of natural tiger habitat, containing extensive forest and wetlands. The region already contains the world's largest single population of tigers (numbering in the hundreds) and offers long-term tiger accommodation. The regional authorities, however, provide thousands of permits annually for access into the Sundarbans, where people carry out logging and fishing; this ultimately results in conflict between tigers and people, leading to numerous man-eating incidents. Between 1975 and 1989, more than 500 people were killed by tigers in this area. More recently, it has been estimated that up to 150 people are killed each year in the Sundarbans of Bangladesh. As a rule, tigers will avoid people; such figures suggest that enforced confrontations by humans in large numbers encroaching the tiger's home range is at the root of the problem. It is clear that the stress of invaders into its core territory is another cause for a tiger to become a threat to man.

Jaguar

General

The jaguar (*Panthera orca*) is the largest wild cat on the American continent. Its range once extended from Texas in the north to Patagonia in the south (Perry, 1965; Mondolfi and Hoogesteijn, 1986). As a result of massive hunting for its coat, the species is now restricted to an area extending from the southern edge of Mexico to the State of Mato Grosso in Brazil. Formerly, it was widely distributed throughout South America (Guggisberg, 1975; Emmons, 1987b). Today, its population has been drastically reduced, limited to three regions: the Llanos (in Colombia and Venezuela), the Amazon and the Pantanal (a region of the State of Mato Grosso in Brazil). The vast Pantanal on the west side of Mato Grosso is an area of about 250,000 km^2, containing abundant wildlife including caiman alligators, collared peccaries and capybaras. The Pantanal is flooded for half of the year and cannot be used for continuous human settlement. The jaguar is a good swimmer and tree climber, abilities that allow it to cope well in this wet environment.

When the region is dry it provides vast areas of grassland. Ranchers now use marginal areas of the Pantanal for grazing cattle, and such animals are very enticing to large male jaguars. When ranchers experience livestock loss, they organize a hunting party to locate and kill the animal responsible. Thus, even in this remote area, the jaguar is not secure.

Appearance

The jaguar bears a strong resemblance to the leopard through its similar coat and general shape. Like the leopard, the jaguar is covered with black spots on a pale yellow background. In some regions, such as the Llanos and the Pantanal, jaguar males can grow to weigh approximately 120 kg and females may be up to 90 kg. These sizes make the jaguar larger than the leopard. Jaguars in Central America, however, only weigh about 60 to 80 kg in regions where prey is not abundant. The jaguar is much more powerful than the leopard and has a larger head, a deeper chest and stockier limbs (Fig. 7.8).

Fig. 7.8. Mature jaguar resting in sternal posture.

The leopard, in contrast, is lither. The coloration pattern of the two species has one slight difference and one marked similarity; while spots cover their coats, the jaguar's spots are larger. In both cases the black spots have a yellow to grey centre, or rosette. As is the case with leopards, a small proportion of jaguar cubs are born completely black. Black jaguars are usually referred to as pumas. Some of the jaguar's subspecies in South America have become extinct through excessive hunting for their pelts. It is remarkable that the jaguar and leopard, which appear to be close relatives, can be so far removed from each other, separated by the world's greatest oceans.

Habitats and feeding

In addition to the principal geographical ranges mentioned above, jaguar populations can be found around river territory in Suriname and the Guianas. Regions in the broad Amazon basin of Brazil support a large population, while small numbers can be found in Bolivia, Peru, Paraguay and northern Argentina. These locations are characterized by long, large rivers, extensive forests of various elevations, tropical or subtropical climates. Tropical forests with streams and rivers provide the favourite habitats for jaguar but some live in distinctly different ecological environments, such as dry forests and swamps (Emmons, 1991). During hot, humid weather, the jaguar enjoys swimming and bathing. Their swimming ability is remarkable, with the help of their huge forepaws (Fig. 7.9) and they will often swim river courses in their travels.

Jaguars can be active on the ground or in trees, and may hunt either at night or during the day. Jaguars will hunt any form of prey that is available, but their principal prey items are the collared peccary and the capybara, which are abundant on the Llanos and in the Pantanal. Depending on its locality, the jaguar also feeds on white-tailed deer, turtles, turtle eggs, caimans, iguanas, armadillos, rabbits and fish. Larger kills are usually relocated to a sheltered place prior to eating. Sometimes, the jaguar will drag heavy carcasses as far as 800 m – even pulling them across rivers by swimming. Depending on the size of the carcass, the jaguar may remain with the kill for 2 days, leaving it exposed and eating it periodically (but never eating decayed meat). Jaguars have differing styles of killing and eating, compared with their larger cousins. Instead of biting the neck of its prey, the jaguar

Fig. 7.9. Jaguar asleep; note size of forepaw.

penetrates the skull with its long canines. The strength in the jaw can even puncture the carapace of a large turtle in the same manner. When the jaguar is feeding, it begins at the forequarters and may leave the hindquarters untouched when it finally abandons the kill. Hunting and feeding are usually performed alone, although mating pairs and mothers with cubs will share their catches. In addition, siblings that are on the move during dispersion may stay together, dividing their kills amongst each other.

Territorial features

The territorial schemes of the jaguar are in the classic feline style. Female ranges are usually located within a male's range and occupy a smaller area. Often, several female territories may be within one male range, without overlapping each other. A pair of females inhabiting the same area would likely be a mother and daughter. In the Pantanal, a female territory was estimated to be about 90 km^2 and a male occupied about 150 km^2. These sizes varied greatly depending on season; in the wet season, for example, they decreased in size. Daily travelling distances were also reduced in the wet season in the Pantanal (Crawshaw and Quigley, 1991). In the Llanos of Venezuela territorial arrangements and sizes were very similar and, evidently, were affected by the wet season. By contrast, in a national park on the border of Brazil and Argentina, jaguars did not occupy exclusive territory; it was observed that four males occupied territories that overlapped extensively (Sunquist and Sunquist, 2002). Population densities of jaguars in various parts of South America vary from about two or three per 100 km^2 in Colombia to about eight per 100 km^2 in Belize. These densities have some relationship to the biomass of the region.

In territorial marking, the jaguar employs the usual methods to indicate its occupancy, principally at the boundaries. Its marking methods include urine spraying on bushes, scraping tree trunks, defecating in noticeable places and vocalizing loudly. In areas where male territories overlap, marking activity is increased. The jaguar's capability for extremely loud roaring is recognized as its principal means of communication. The roar takes the form of rapid repetitions of a guttural sound of great volume. Both males and females participate in roaring sessions, some of which can be made between two individuals roaring and counter-roaring at each other over a distance. Such sessions of mutual vocal exchanges can last for hours and clearly serve as socio-sexual communication. In addition to advertising an individual's location, such calls serve to bring males to females for breeding, since jaguars usually exist alone within the boundaries of their home territories.

Breeding

Adult jaguars live alone, associating with one another only during the short breeding phase. It has been observed (Almeida, 1986) that a female jaguar in heat will travel frequently, making calls that attract male jaguars, leading them to follow her prior to mating. The female may remain in heat for as long as 17 days, but periods of about 6 days are most common. The oestrus display includes restlessness, rolling, pacing, rubbing against the attending male and making caterwauling sounds. During copulation the male crouches over the female, gripping her neck in a bite without resting on top of her. The female's lordosis posture aids the male in achieving intromission. This is followed by pelvic thrusting, which is continued briefly before dismounting. At that moment the female utters a screech and turns on the male, attempting to bite him. Following a brief interval, copulation is repeated with the same female reaction. Jaguar pairs may copulate about a hundred times per day (Lanier and Dewsbury, 1976; Eaton, 1978; Seager and Demorest, 1986).

Following mating, the male leaves the female to raise her cubs on her own. Cubs are born after a gestation period of about 100 days and litter sizes may number up to four. Births occur at any time of year in warmer regions, but more often during summer months in temperate regions (Sunquist and Sunquist, 2002). The birth site is usually within dense cover in a secluded spot. At birth the cubs have thick, woolly, spotted coats. Their eyes open any time between 3 and 12 days of age and they suckle their mother for at least 13 weeks. Once the cubs have reached this age they begin to eat meat supplied to them by their mother. At 2 months of age they begin accompanying her on hunting trips, although they will often continue to suckle until they are nearly 6 months old. This is their hunting tuition, and in these initial stages they spend their time observing. As the cubs get older, they begin feeding from the catch. They grow rapidly, with males gaining 25% over females at 1 year of age and 50% when they are 2 years old. This is the common difference in growth rate among infant felines in general.

Development

During their main period of growth, young jaguars spend most of their time in close association with the mother. At 1.5 years of age they begin to hunt for themselves within the natal range. As they approach their second year of age they are now independent individuals, and the process of dispersal has begun. Males travel away from the natal area as far as 30 km (Sunquist and Sunquist, 2002), while young females wander shorter distances and may even return to the maternal range. Males may continue searching for a new territory for months, facing a variety of hazards on their journey. It is estimated that a high mortality rate among male jaguars during their dispersion comes as a result of being hunted. In the wild, jaguars may live to be less than 11 years old. In captivity, they adapt well to their circumstances and many live to be over 20 years of age.

Welfare

Between 40 and 50 years ago, as many as 10,000 jaguars were killed each year for their skins (de Almeida, 1986). In the last several decades, jaguars have been listed as a protected species and the skin trade has declined sharply. Poaching continues, however, and while cattle ranchers also contribute to the killed numbers, the principal threat to the species appears to derive from loss of habitat due to human encroachment on their ancient territories. Increasingly, forestland is being cleared away to create grazing sites for cattle, bringing the jaguar in closer contact with livestock. Despite close proximity, not all jaguars are tempted to hunt cattle. One study found that those animals responsible for such killing had been injured from shooting, and as a result were less able to hunt their natural, wild prey (Rabinowitz and Nottingham, 1986). The current number of jaguars has been greatly reduced all over South America. Even in the Pantanal, the most protected area, it is estimated that only about 1500 jaguars remain (Crawshaw and Quigley, 1991; Quigley and Crawshaw, 1992). This animal's future is in doubt in spite of its great physical attraction.

The total number of jaguars remaining in the world now is estimated to be between 12,000 and 17,000. This compares with the population in 1950 which was 400,000. The current numbers are distributed throughout the whole continent of South America and countries of Central America. Their surest habitats are in rainforests, particularly those in Brazil's Amazon (with over 4,000,000 km^2), the Maya Forest in Belize, Mexico and Guatemala (with 52,000 km^2) and the Moskitia Forest in Honduras and Nicaragua (with 24,000 km^2). Currently, there are jaguars in the following countries: Argentina, Belize, Bolivia, Brazil, Colombia, Costa Rica, Ecuador, French Guiana, Guatemala, Guyana, Honduras, Mexico, Nicaragua, Panama, Paraguay, Peru, Suriname and Venezuela. This magnificent feline of the American continent has a doubtful future.

8 Principal Species of Wild Cats

In order to appreciate the domestic cat's inherent nature, it is useful to compare it with other family members living in the world's wilds. Felines are all related through a prehistoric common origin. They retain some ancient hard-wired neural networks, while acquiring some new genes that give variety among the various species in this remarkable family of beautiful carnivores. In addition to the domestic cat (*Felis catus*), the extended feline family consists of 38 other cat species worldwide. While these are wild-living and wild-hunting, the housecat shares many of their traits in modified forms. No other animal family matches felines for their lithe style, colourful coats, athletic stealth, periodic lethargy, intense action, ferocity, controlled predation and pure beauty. The purpose of the latter defies explanation except camouflage, but it makes the wild species the target of human predation. Intense action is a need in predation and the energy expenditure in hunting, capturing, killing and carcass dissection normally limits the quantity of kills to ingestive need. The carnivorous meals provide a concentration of nutrients and their digestion calls for a period of physical inactivity. Hunting, with its work load and the ample food obtained, is governed by hunger. Some features of wild felines are included in the ethos of the domestic cat.

Typical members of the cat family share a number of physical and functional features such as the following: good piercing tusks, exceptional eyesight, long tails, upright ears, firm whiskers, rotatory carpal joints, broad strips of apocrine glands in the area of the temples, retractable claws, high hard palates, straight rows of incisors that place the tusks in a forward position and well-haired ear openings.

Leopard

The leopard (*Panthera pardus*) is a beautiful cat in the genus *Panthera*. Its spots are of various designs and cover its entire body, head and tail. Some leopards have a solid black colour created by a recessive gene and are called panthers; their spots, while present, are faint. The leopard's beautiful coat made it a target for hunters supplying the furrier trade. The trafficking of leopard skins is now under some control. In former times, about 5000 leopards were killed annually for their pelts.

Leopards have a wide distribution, ranging over much of Africa and Asia. They number about half a million and are more numerous than the total population of all the large cats in the world. Within their locations, males in particular have very well-defined territories in which they live alone, except during breeding times. Their territories are often about 25 km^2 and they patrol their territories constantly. They mark their territories in the normal feline style by spraying on to bushes and trees, making long scratch marks down the trunks of trees. They leave their faeces uncovered as territorial marks but cover their faeces in the core area of the habitat. Both male and female leopards perform this marking of territory, but males are more active in marking their territory from which they exclude other males aggressively. The leopard's territories are commonly in forested areas, but they also live successfully in a wide variety of other habitats such as semi-desert regions and high altitudes of mountainous places, to the limits of tree growth (Batley, 1993).

The male leopard averages about 67 kg in weight, while females are about half that size. Asian leopards are smaller than those in Africa. The males only consort with females during the short oestrus periods of about a week. If she is without a litter, the female may breed during any time of the year. The mothers keep their young with them for about 2 years, and hunt for their food until they have learned the techniques of successful predation, which is an adult's business. In the catching and killing of large prey, there is either immediate success or failure – sprinting energy for a second try is not available. The males have no role in the care of cubs, which are very small at birth. The mother leopard, like the domestic cat with newborn kittens, carries her young to new hiding places every few days until they are several weeks old. The leopard cubs open their eyes at about 10 days of age, when they are about 1.3–1.8 kg in weight. Commonly there are three cubs per litter. Occasionally one litter may contain a black cub in addition to spotted ones, proving that the panther is actually a leopard (Sunquist and Sunquist, 2002).

The richly spotted leopard has an excellent camouflage in the bushland it favours. It is renowned for its ability to avoid detection. This allows it to get close to prey before pouncing. It hunts at dawn or dusk after taking up its lurking position in thick bush or from a tree. The leopard is an extremely competent tree climber and has the habit of carrying its catch high up into a tree before eating. Its common prey includes young wildebeest, impala and other small antelopes, warthogs, young baboons and any small animals or birds in its territory. It will also eat snakes, lizards, hares, hyrax and rats. Although this animal usually drinks on a daily basis, it is known to be capable of living for a week or more without water, particularly if there is a need to hide from the near presence of lions. Particularly in India the leopard may encroach on human communities, catching cats or dogs and scavenging from refuse. They are not in any way welcome in such activities, but they are surprisingly tolerated as pests.

The leopard is particularly secretive and constantly tries to avoid confrontation with its natural enemies that include the lion, the hyena and the baboon. When it has young cubs it is extremely protective of them since they are targeted as easy prey by hyenas and even male leopards. The latter will readily kill cubs to deprive the mother of her maternal status and cause her to return to oestrus, making her into a breeding mate for the cannibalistic male leopard. In oestrus, the female leopard makes loud, repeated vocalizations and goes through the usual feline signs of oestrus such as rolling on the ground. She usually mates with more than one male.

Snow Leopard

The snow leopard (*Panthera uncia*) is a different species from the common leopard. It inhabits northern India and the lands that border the Himalayan mountain range. Populations often exist in geographical pockets largely in China, but also in Russia and Mongolia. Its high-altitude environment does not contain a plentiful supply of prey, and it is often forced to travel great distances in search of food (Jackson and Ahlborn, 1989).

The diet of this species consists primarily of blue sheep and ibex, but other animals such as wild ass, musk deer, wild pigs and gazelles are also preyed upon. Small prey, including marmots, hares, rabbits and partridge are also hunted, and, unlike other cats, this feline will eat carrion. Another strange feature of the snow leopard's diet is the quantity of vegetation it consumes. All cats are known to eat small amounts of greenery, but no other ingests an equivalent quantity to the snow leopard – nearly half of their excrement has been found to contain vegetation, and some deposits are composed entirely of grass and shrub matter. This cat will also target livestock, and for this reason, its presence is in jeopardy in such areas as Anapurna.

The snow leopard varies in size, up to 67 kg in weight and has a beautiful spotted coat. The larger spots are dark rings that are arranged in distinct rows on a grey or yellow background which grows into a paler shade in winter, giving this animal a summer and winter coat. Its beautiful pelt made it a target for hunters, reducing its population and placing it on the endangered species list. The hunting of these spectacular animals is now prohibited, yet they are still targeted by poachers (McMahan, 1986; Broad *et al.*, 1988).

Snow leopards have a number of adaptations, including high-set eyes (which allow them to see well from a low crouched position), thickly haired paws (which provide insulation for the pads in winter) and very dense fur (for bodily insulation). Their powerful hindquarters fuel their spectacular ability for upward leaps into the rocky hillsides of their habitat (Wharton and Freeman, 1988).

This wild cat is fairly unique in its distinct breeding season. The female comes into oestrus when winter is over, and one or two more oestrus cycles follow until pregnant. Males will travel great distances during this time. Cubs are born in May in a secure den, lined with the female's own belly fur for insulation. The kittens have closed eyes until about 10 days of age, at which point they begin to crawl about. They continue nursing from their mother until they reach 2 months, and by this time can consume pieces of flesh. Litter sizes often consist of four or five individuals that mature rapidly. The cubs remain with their mothers throughout their first winter, learning how to hunt and cope with their particular circumstances. In the typical feline style, the adult snow leopard lives a solitary life in a home range that may vary from 14 to 140 km^2. Within its chosen individual territory, this cat can live up to 20 years.

A safe reserve for the snow leopard has been established in Bhutan, central Asia. Many of these cats are now well established in captivity where they breed with much success. The principal dangers to their welfare are the loss of their natural prey, conflict with livestock owners and poaching for the illegal marketing of their skins.

Clouded Leopard

Even as a smaller member of the family of wild felines, the clouded leopard (*Neofelis nebulosa*), or 'tree tiger', is a species that is a formidable hunter. The male's weight is 125 kg on average, while the females are considerably smaller. Among its notable features are: a long tail extending about 1 m in length, a blunt snout with long tusks and a coat with a broken patchwork design featuring spots and stripes on a background of greyish yellow. The blunt snout facilitates a very typical feline dental feature. With a short nose and a straight row of small incisor teeth, the tusks or canines are placed in a forward position, close to the front of the mouth. Feline canines are like fangs; strong, slender and sharply pointed. This feature is of great advantage in catching prey, so much so that if a wild feline breaks a tusk it is likely to die of starvation, having lost its essential equipment for clamping on to prey.

The clouded leopard has a remarkable world distribution as a species. Although its population has been greatly reduced by hunting, it still exists south of the Himalayas into South-east Asia, down into the Malaysian archipelago and Indonesian islands. Other countries included in its distribution are Nepal, Thailand, Vietnam, south China, Sumatra, Borneo, Java and Taiwan. Clouded leopards are regarded as tree tigers quite appropriately, as their favourite place is reclining along a tree branch. Their camouflaged pelt allows them to blend in with surrounding foliage. The usual home territory is dense forest. With the severe, clear-cutting of trees in many areas of South-east Asia, the clouded leopard has lost much of its natural territory, which contributes to its scarcity. This cat displays excellent balance with its long tail running along the back of a tree. It is a most notable tree-dwelling cat and another feline listed as an endangered species with international protection. Despite protection, poaching continues to reduce species numbers.

The clouded leopard creates a ground den and breeds once a year. The litter usually contains two cubs. The cubs have closed eyes for 12 days and then become increasingly active and able to climb once they are 1 month old. Although they can eat flesh at this age, they are not weaned until they are about 5 months of age, by which time they are well grown. At birth they lack the spots and coloration of the adult. Once weaning has occurred they will acquire the adult coat pattern. In addition to large patches of grey with black borders that cover the animal's trunk, stripes run down the length of the back while the tail has black rings. Black spots and stripes cover the head. These features on a grey to yellow background give the animal a very striking appearance.

In true feline fashion, the male lives a solitary existence in its own territory, except at mating time, which may last for a week to 10 days. In the care of its home range it is not very aggressive or fierce. The female has the company of her cubs until they are young adults and choose to wander away in search of their own territories. The vocalizations of the clouded leopard include purring, but not roaring. For some biologists, their long tusks are reminiscent of the long-extinct sabre-toothed tiger, from which they believe the clouded leopard is descended. Such a claim could, of course, be made for all other cats, as they are clearly related. The miniature household variety of cat, for example, may have descended from the huge sabre-toothed tiger of the Ice Age.

The tree tiger's principal source of food is other tree inhabitants, namely birds and monkeys. It also preys on deer, goats, wild pigs and young cattle, and is a crepuscular hunter when on the ground, limiting its hunting to dusk and dawn. Many house cats have peaks of activity in early night and morning, suggesting that there is a remnant of crepuscular action in their genotype.

Recently a new species of clouded leopard has been identified in eastern Asia (BBC Earth News, 2011). It is called the sunda clouded leopard (*Neofelis diardi*) and has two forms: one in Borneo and the other in Sumatra (*N. diardi bornensis* and

N. diardi diardi, respectively). These leopards live in dense jungle and have very secretive lifestyles in trees. The jungle on both islands contains a rich variety of prey such as birds and monkeys.

Ocelot

The ocelot (*Leopardus pardalis*) lives on the American continent and is a medium-sized wildcat, weighing 8–10 kg on average (Gardner, 1971). Its coloration is similar to the jaguar and leopard and it is sometimes called the 'dwarf leopard'. It has a deep yellow or orange coat with black stripes, spots and rosettes. The coat is sleek and white underneath. The tail has black bands and the back has irregular markings, while the back of each rounded ear has a large circular white area called ocelli. In addition, black stripes run along the side of the face and over the skull to the back of the head, bringing attention to its large eyes. The ocelot is similar physically to the nine other species or subspecies of *L. pardalis* that occupy the same territory from the Amazon to Texas. They constitute the dainty cats of the *Leopardus* genus. The ocelot is the largest of these cats, and has been hunted extensively for its coat. It is estimated that hundreds of thousands of ocelots were once killed for their pelts and became rare until protected. Their numbers are now showing some increase according to the International Union for Conservation of Nature and Natural Resources (IUCN) Red List of 2008 (Wikipedia, 2010).

The ocelot's main American territory is the northern tropical belt between the Equator and the Tropic of Capricorn, which extends from the Amazon to Mexico. Its common name varies among the numerous countries included in this range. It is therefore a warm weather cat, as is the domesticated cat. The ocelot's home range varies in size according to food availability and density of cover (Ludlow and Sunquist, 1987). Males may use up to 50 km^2, while females use about 15 km^2 and they identify their individual territories by spraying urine and depositing faeces in very obvious places, sometimes excreting frequently at the same location. The male ocelot will defend its own territory fiercely against another male of his own kind and the female also keeps its territory free of competition. The usual habitat is in an area of dense vegetation in tropical forests and mangrove swamps. Ocelots hunt at night (Emmons, 1987a, 1988; Emmons *et al.*, 1989); at this time they may travel across open land. During the day the ocelot often occupies a tree, lying along a branch watchfully. During a hunt they will utilize their keen sense of smell and sight to seek out small forms of life such as rodents, frogs, crab, turtles, fish, birds and rabbits. Large male ocelots will tackle deer and opossums (Sunquist, 1992).

Ocelots breed once every second year. The male remains with the female during her oestrus period, which lasts for over a week. Once mating is complete, the male leaves while the female constructs a den in a concealed place, such as a cave or a dense thicket. The ocelot is not a prolific breeder. At birth there is usually only a single kitten. The kitten has a thin covering of hair and its eyes are closed for 2.5 weeks. At about 3 months of age the kitten becomes weaned and starts to wander out of the den. It remains with its mother for 2 years before leaving her to establish its own territory (Crawshaw and Quigley, 1989).

Increasing numbers of ocelots are being sought in the USA by people intending to keep them as domestic pets. This practice is viewed with differing opinions. Large numbers are also housed in zoos, which seems to be the trend with every species of wild cat from around the world. Certain wild feline species are more numerous in captivity than in the wild. The zoo environment, however, protects them from poachers, while creating a safe haven for much successful breeding.

African Golden Cat

Another medium-sized cat that exists in Africa is the African golden cat (*Profelis aurata*), which can be found within the rainforest regions of West and Central Africa. It is the only member of its genus, but a close relative to the caracal and serval cats. The golden cat male weighs between 9 and 16 kg, while females are slightly smaller, measuring 60–100 cm in length with a short tail between 17 and 35 cm long. They are therefore about twice the size of the typical housecat. Its body colour is variable, with colours ranging from golden cinnamon or reddish brown to grey or black. Some individuals lack spots, while others have spots that range in colour from tan to black. The face, throat and undersides are of a light colour, even appearing white. The tail has bands of various colours and the tip is always black. The body colour has two phases; a golden to red shade in one phase and grey in the other, these phases last for months and then alternate. The fur has a very notable identification feature – a ridge of upstanding hair in front of the shoulder.

The African golden cat is strongly built with long, powerful legs and large paws, making it capable of coping with its dense jungle habitat (Brugiere, 2001). This habitat allows the cat to live a very reclusive life. Its active hunting time is night, chiefly early evening and pre-dawn. It feeds on rodents, small monkeys and pigs, and, when in close proximity to a human settlement, it preys on small livestock.

This animal is a regular breeder. The litter usually contains two kittens that open their eyes during their first week of life. The kittens develop rapidly and are weaned before they are 2 months old. Females attain breeding ability at about 1 year of age while males do not breed until they reach about 1.5 years. It is estimated that this animal's lifespan is about 20 years.

Caracal

Despite its substantial population in Africa and the Middle East, the caracal (*Caracal caracal*) is seldom seen as it is particularly secretive and small in size. Its physical characteristics include elongated, tufted black ears, which give it a lynx-like appearance. For this reason it has been called the 'Persian lynx', 'Egyptian lynx' or 'African lynx' depending on its location, even though it is a different species most closely related to the African golden cat and the serval. While classified as one of the small wild cats, it is one of the heaviest, possessing long legs and great strength. On average males weigh between 14 and 18 kg, while females' average weight is 11 kg. Males measure just under 1 m in length, while females are shorter but have a similar general appearance. Both sexes have short tails (Stuart, 1986).

Caracals are slender, muscular and capable of great speed and spectacular jumping ability. They live on the ground in sandy desert, semi-desert, dry steppes, savannah and scrub forest. Their grey, pale red or sand-shade coat coloration provides them with exceptional camouflage. An alternative colour is black, which suits their nocturnal habits. Thick tufts of stiff hair between the pads of their paws facilitate a good grip on sand. The caracal travels on open country where there is partial rock cover and bushes suited for hiding. Hiding and pouncing is utilized as a tactic for hunting (many housecats practise spells of hiding and pouncing). A caracal's leaping ability makes it adept at catching birds, which make up a proportion of its diet, together with hyraxes, the common 'rock rabbit'. In addition, it is not uncommon for a caracal to tackle larger prey such as small antelope. Any uneaten portion of the carcass is hidden by scraping surrounding soil to conceal it, after which the cat will return later to finish it. Caracals can survive without water for extended periods deriving fluid from the tissues of prey, like the housecat obtaining moisture from soft food.

Breeding takes place during any time of year when food is adequate for the maintenance of good health. It is therefore welfare dependent. Oestrus continues for 2 weeks. The female caracal performs urine spraying around her territory during her oestrus period and this attracts male caracals, which fight for breeding rights. The female will usually mate with several different males, and some of these males will remain with her for many days during this time. This is an example of natural selection at work. Litter sizes are larger than most other cats, numbering five or six kittens. Their eyes open at 10 days of age and they are able to crawl out of the birthing den at 1 month. At this stage the mother carriers her kittens to a new den and periodically takes them to different locations. Caracals have a life expectancy extending through their teens (Wikipedia, 2010).

African farmers regard caracals as pests since they are known to prey on small livestock. For this reason these cats are targeted by hunters. Unlike other felines, their fur has no commercial value. Unaggressive in nature, individuals may be acquired as pets when they readily adapt to a domesticated existence, accepting human companionship without mistrust. It has been reported that scientists at Moscow zoo successfully crossed a caracal with a domestic cat, resulting in the birth of hybrid kittens. The housecat's genes may not be that far removed from those of its wild relatives, however ancient the connection may be now.

Serval

The serval (*Leptailurus serval*) is another African wild cat that is genetically linked to the African golden cat, although it differs in appearance with flowing lines of black spots running across its tawny-coloured body. A few stripes run down its neck and its legs are banded. Black servals are common in some parts of its African range.

These are often considered to be black panthers, the melanistic leopards.

In addition to the golden cat, DNA studies find the serval to be very closely related to the caracal and the cheetah. It is estimated that the cheetah is actually a descendent of the ancient serval (Kingdon, 1977; Wikipedia, 2010). In some behavioural ways the serval is a mirror image of the domestic cat – it can be house-trained, for example. Some cat fanciers consider the serval crossed with the domestic cat to be highly appreciated as a pet, as illustrated through interbreeding. Lousada (1956) and Leyhausen (1988) have shown that such cats have a degree of inherent sociability. This factor has also been demonstrated in the serval by Smithers (1978, 1983).

Serval females weigh between 7 and 12 kg, while males average 9 to 18 kg. Its physical features include long legs and toes. It can sprint, swim and climb trees. Its territory does not include deserts, semi-deserts, dry steppes and dense jungle, but always contains a river or stream. A very efficient hunter, the serval can run at great speed and perform spectacular leaps. It hunts predominantly at night and preys on rodents, birds, hares, hyraxes, frogs and fish. With its tall, rounded ears it exhibits a great sense of hearing, detecting the sounds of underground rodents. It digs these out and swallows them whole (Wikipedia, 2010).

When prey is plentiful the serval does not have the need to travel far. As with most cats, the serval leads a solitary life in its range. The female's range extends between 10 and 20 km^2. Males have larger ranges, up to 30 km^2. They mark their home ranges by spraying urine on to the ground or over bushes, which they then scratch with their claws (Bowland, 1990).

Living in tropical Africa, there is no seasonal breeding. The female may come into oestrus at any time of the year. Her period of heat usually lasts about 4 days, during which time numerous matings occur. The litter, consisting of two or three kittens, is not born in a den but rather in any suitable sheltered location, such as beneath a bush. The kittens open their eyes between 10 and 14 days of age and can eat solid food when they are about 1 month old. Weaning follows by 2 months of age. At 6 months their permanent tusks have grown, allowing them to hunt for themselves. The young servals will remain with their mother for a year. In the wild, they have a life expectancy of about 10 years, but in human care that expands up to 20 years.

Historically, serval cats have been kept as pets on different parts of the African continent. The ancient Egyptians worshipped their serval felines as gods. In recent times, they have been acquired as pets in Europe and North America, having been found to be easy to domesticate and dedicated to their first owners. The serval crossed with the domestic cat creates hybrid offspring, termed Savannah cats. These are smaller than the serval but retain its colour and markings. They are also more sociable and highly tolerant of strangers, making them easy to train. They are, of course, susceptible to all the common diseases that can affect housecats. Savannahs are not regarded as wild cats, but as domestic animals and are registered with the International Cat Association.

Cheetah

Being a specialized animal built for phenomenal ground speed, the cheetah (*Acinonyx jubatus*) differs from most other cats. It is not a tree climber, but can sprint for distances up to 460 m at speeds ranging from 112 to 120 km/h. It is similar in size to the leopard, with adults weighing from 38 to 60 kg. The males and females are of equal size and the sexes are difficult to recognize on the basis of body size alone. Its full body colour is a shade of tan, thickly covered with dark spots. It has a deep chest, a slender waist, long legs, feet with slightly webbed toes, extendable claws without sheaths and a smallish head that adds to its streamlined appearance. The animal has characteristic 'tear marks' or dark lines that run from the inner angle of each eye to the corner of the mouth on the same side (Caro, 1994).

The species displays some variety in colour; the so-called 'king cheetah' has blotched marks over its body and broad lines running down its back. Rare coloration includes grey, white and black (melanistic). Some desert-dwelling individuals may be very pale in colour.

The cheetah has an extensive geographical range, although it has been eliminated in countries such as India, as a result of being hunted for its pelt. Cheetahs once existed in 44 countries in Africa but are now found in just 29. Its present concentration is in southern Africa, but a dispersed population occurs throughout the rest of Africa and West Asia. The Khorasan province of Iran is home to a rare species of cheetah (numbering about 50–60 individuals) that has now been found to belong to a separate subspecies (*A. jubatus venaticus*). These are, in fact, the last remaining survivors of an ancient Asian cheetah. Six different subspecies of cheetah are currently recognized. All cheetah subspecies are now considered to have descended from an ancestor that existed 11 million years ago (Sunquist and Sunquist, 2002).

The female is able to breed at about 2 to 3 years of age. Regarded as promiscuous, they have been observed breeding with different males in one heat period, but this is a feline characteristic that can be seen in the common housecat. The average litter size is about five cubs. Unfortunately many cubs are killed by intruding predators. Those surviving develop woolly manes called a mantle which is later shed. They develop quickly and become weaned by about 2 to 3 months. These grown cubs stay with their mother until they are between 1 and 2 years old. By this age they have been taught to hunt, and their mother leaves them on their own. The cubs remain together as a sibling group for a further 6 months or so. At about 2 years old the females leave the group but the males stay together and remain as a social group for life. These male groups are called coalitions. Some males do exist alone, however (Caro, 1994).

In their social system the female lives a solitary life when she is without offspring. The male coalitions acquire their own territory that they maintain for about 4 years. The territory they choose invariably overlaps the territories of several females, with which they breed. A coalition's territory may extend to 160 km^2. The males mark their territory by urinating on prominent objects such as trees and mounds, with all males in the coalition contributing to this scenting procedure. These males will attack any animal that intrudes, resulting in fighting and even death.

Cheetahs live in various habitats, ranging from semi-desert to thick grassland, savannahs, dense vegetation and mountainous lands. They use expansive areas when hunting, preying upon springbok, impala, gazelle, guinea fowl and hares. Their vision is an essential tool for hunting. Unlike other felines, which are nocturnal in their hunting habits, cheetahs are particularly active in early morning and evening. The physical speed of the cheetah during a hunt causes a dangerous rise in its body heat, and the animal will usually rest following a kill, prior to eating it. If there is a risk of another predator, however, it will eat the carcass quickly. In such circumstances, the cheetah is outranked by such threats as lions, leopards and hyenas and does not contest its kill. If the prey escapes at the end of a chase the cheetah will not make a second attempt. The speed of this feline, however, allows it to outrun all other animals, giving it the advantage of escape from predators.

Normally cheetahs do not prey on domestic livestock, but many farmers were formerly of the belief that they did and for this reason they were often hunted down. The cheetah was also hunted for its pelt, but this is now under some control. Illegal hunting still occurs, however. Some of these cats were also captured to be kept as pets by the ancient Egyptians. The cheetah is not an aggressive animal, and its capture has now been made illegal. It has long been recognized that its social behaviour changes under environmental pressure (McVittie, 1979). This species has been endangered for some time (Eaton, 1974) and cheetahs are now listed by the IUCN as a vulnerable species. Approximately 12,400 cheetahs currently remain in the wild. Attempts are being made to breed them in captivity in order to restore their numbers in areas where they once thrived and, fortunately, there has been some success.

Bobcat

The bobcat (*Lynx rufus*) is the most common of the North American wild cats (López-González *et al.*, 1998). Roughly the size of a large dog, it has a short tail and russet-coloured coat with a mixture of black stripes and spots over its entire body. Its condition varies considerably with the availability of prey (Matlacak and Evans, 1992). Distribution is limited to the North American continent, where it lives a solitary, secretive lifestyle in a variety of habitats from mountainous lands to prairies while avoiding open areas (Delibes *et al.*, 1997). Males may have territories of up to 64 km^2. Females have smaller territories and these are often within a male territory. Bobcats mark their territory by spraying urine in various places and depositing faeces on prominent locations. In addition, they scrape soil in patches, digging it up.

The bobcat travels only at night, often covering great distances (Koehler and Hornocker, 1991). Another crepuscular cat, it hunts primarily during dawn and dusk. Its food varies with the seasons (Matlacak and Evans, 1992). Hares and rabbits constitute its principal diet; however, this cat has been known to prey on small domestic livestock such as sheep and goats when hares are scarce. Its manner of catching prey is similar to all cats. Creeping as close as possible to the unsuspecting kill, it then pounces, seizing the prey by the neck (Delibes *et al.*, 1997).

The bobcat breeds in the winter every other year. The bobcat female makes or finds a concealed den for the birth, and has been reported as choosing an abandoned beaver lodge for that purpose (Lovallo *et al.*, 1993). When the cubs are born, the female drives the male away. He returns to his own territory, but returns periodically with food for the cubs. The newborns open their eyes in 1 week and continue to feed from their mother for 2 months. Once they are weaned the male will return with food. As the cubs grow, they travel with their mother when she hunts throughout her territory, remaining with her for about 9 months before they leave to find territories of their own. Bobcats are hunted for sport and trapped for their pelts.

Lynx

General

Most feline species have specific geographical homelands even though many of these are very expansive. The lynx is a notable exception, having a distribution which spreads over the northern hemisphere. One reason for their dispersal is the fact that there are four different species of lynx, namely the Eurasian lynx (*Lynx lynx*), the Canada lynx (*Lynx canadensis*), the Iberian lynx (*Lynx pardinus*) and the bobcat (*Lynx rufus*). The caracal (*Caracal caracal*) is commonly called the 'Persian lynx' or 'African lynx', but is classified in a different genus (Sunquist and Sunquist, 2002).

Canada lynx

The Canada lynx (*L. canadensis*) is of medium size – about 10kg on average. Its coat colour is a uniform mixture of grey and black mixed with buff, while its underside is mottled. The tail is short and has a completely black tip, distinguishing it from the bobcat. The tips of the ears have prominent tufts of black fur characteristic of all lynxes. The paws of the Canada lynx consist of long, dense fur, creating a snowshoe resemblance. In heavy snow these furry paws can spread widely and allow the animal to grip the surface as it travels (Breitenmoser *et al.*, 1993).

The distribution of the Canada lynx extends across the northern broad belt of boreal forest that stretches from Alaska in the west to Newfoundland in the east. A small number of these lynxes can also be found in northern parts of the USA. Its distribution matches the geographical spread of the snowshoe hare, its principal prey (Poole, 1994, 1995, 1997). This animal does not inhabit the plains of Manitoba or Saskatchewan as they are devoid of natural cover. In extremely cold weather, the Canada lynx will seek shelter in the depths of forests where there are thickets of bush, tangled branches and stumps (Matlacak and Evans, 1992).

When snowshoe hares are scarce, the lynx population drops dramatically (Brand and Keith, 1979; Murray *et al.*, 1994). Although this lynx may travel during the day, it is a nocturnal hunter. Seeking out hares by sound, it consumes on average one hare per day. The Canada lynx will often hunt in small groups. A mother accompanied by her kittens is particularly successful in capturing prey, compared with lone males. It is not uncommon for gatherings of these lynxes to engage in communal play (Sunquist and Sunquist, 2002). Within their territories, adult lynxes mark their boundaries by spraying urine and depositing faeces, while kittens bury their excrement. Scent-marking is also performed by rubbing their cheeks on prominent objects.

The Canada lynx breeds once a year and the female remains in oestrus for about 3 to 5 days. While litter sizes vary from one to eight kittens, kitten mortality is often high due to the mother's lack of milk – a result of her own hunger. In Newfoundland, most lynx kittens are born near the end of May, when better weather is established. Well-fed kittens grow fast and their eyes are open by 2 weeks of age. By 3 months of age the kittens are weaned. At 7 months they accompany their mother as she hunts and remain in her care until they reach about 10 months old, at which age they disperse. Some dispersals extend to hundreds of kilometres. Residence in the new territory may last the life of the individual. In the wild, some of these lynxes are known to live up to 14 years.

Eurasian lynx

A century ago, no other cat occupied such an expansive area of the world as did the Eurasian lynx (*L. lynx*). Its distribution once covered much of northern Europe, Asia and China. Although its territory is now reduced, it remains immense, extending from Norway, Sweden and Finland to Poland, Russia, Siberia, Manchuria, Mongolia and northern China. This lynx has a very ancient lineage that goes back 10 million years.

The Eurasian lynx possesses a general resemblance to other lynxes and the bobcat. The largest cat of the group, it is roughly twice the size of the Canada lynx. Its long hindlegs, while appearing leggy, are thick and powerful. It has a short back, stumpy tail and the characteristic tuft of black hairs extending from the tips of its ears. Another special feature is webbing between its toes and a mass of hair covering its feet. Its dense fur coat is of greyish appearance and a thick ruff of hair around its neck and face creates the appearance of a mane. Its overall coat is long silver grey fur with a silky character. The undersides of its body are white. Males are usually much larger and heavier than females, with individuals weighing roughly 40 kg.

This cat chooses to live in deciduous and boreal forests (Okarma *et al.*, 1997). When food is scarce, however, it can easily adapt to other habitats and will migrate extensively in search of prey. Generally, this lynx is a very cautious cat and an excellent tree climber when it chooses to conceal itself. During the day it hides in rocky caverns and lies in deep forest bush, using fallen branches and loose roots as cover.

The Eurasian lynx is another cat with a crepuscular timetable for its hunting excursions, often travelling throughout its territory all night. On rare occasions it ventures abroad in daytime and may find its way into villages. It commonly seeks out roe deer, reindeer and other young ungulates; hares, wild pigs, beavers, rodents and marmots are also preyed upon. Although the Eurasian lynx primarily hunts alone, it may also hunt in a pack of several individuals. During a hunt the pack may travel great distances, often covering 10 km in one session. Eurasian lynxes are known to kill large numbers of prey in unusual circumstances, such as a herd of semi-domestic reindeer. A normal kill consists of about one deer per week. Hunting proceeds with creeping and springing in the usual feline fashion.

Male lynxes may roam outside their territories during the mating season. They will remain in their home range, however, to mate with females that have established their territories within. Females in oestrus make loud calling sounds, particularly at night. Oestrus lasts about 4 to 7 days, during which time the animal displays the oestrous state by rolling and meowing blatantly. Repeated matings occur within this time. The pregnant lynx will then create a birthing den to house her litter, consisting of one to four kittens.

The kittens open their eyes at 10 days of age and begin consuming solid food when they reach about 6 weeks. At 3 months, they accompany their mother when she abandons the den. Weaning does not occur, however, until they reach about 6 months of age. The young lynxes will continue to travel with their mother until they are 10 months old. Once the next breeding season has commenced, these young lynxes will go in search of their own territories. The lifespan of the Eurasian lynx is known to extend to 21 years in captivity. Lynxes were once hunted for their fur, but are now given legal protection in most European countries.

Iberian lynx

The Iberian lynx (*L. pardinus*) was once a well-established species, with a population of thousands in Spain and Portugal. Today, its numbers are reduced to several hundred, with an estimated 800 individuals in south central Spain (Ferreras *et al.*, 1992; Beltrán and Delibes, 1993). Problems began in the 1950s when the myxomatosis virus was introduced to Europe. This virus destroyed large numbers of rabbits, which comprised the main food source for the Iberian lynx. As a result, lynx numbers fell dramatically from starvation. In addition, a new agricultural plan deprived the lynx of much of its habitat. The population became further challenged with an outbreak of canine distemper that had been acquired from farmers' dogs.

The physical features of this lynx include a short body and tail, long legs, smallish head and tufted ears. A medium-sized cat with a coarse short coat, its colour may be bright yellow, red or tawny with dark spots. The spots are arranged in parallel rows from the shoulders to the flanks. The undersides of this cat are white.

The Iberian lynx inhabits areas frequented by rabbits, their principal prey (Beltrán *et al.*, 1996). Primarily a nocturnal hunter, this cat is generally observed in the morning and evening when it returns from or commences a hunt. While rabbits are its predominant food, it will also prey upon deer, ducks, geese, rodents, snakes and lizards (Gaona *et al.*, 1998). When catching rabbits, a game of catch and release is played, allowing the rabbit to escape momentarily – in similar fashion to the domestic cat with a mouse. When they catch a deer, they will drag the carcass to a secluded area to consume it (Ferreras *et al.*, 1997).

The breeding season begins in the months of January and February. Litters, which usually consist of two kittens, are born in March and April. The kittens are weaned gradually from 2 months of age and remain with their mother for 1 year before dispersing in search of their own territories. Unfortunately, they face a high mortality rate from human contact, due to traps and road accidents. Their natural lifespan is about 14 years. The population is declining rapidly (ICONA, 1990; Palomares *et al.*, 1991; Rodrígez and Delibes, 1992).

Geoffroy's Cat

The Geoffroy's cat (*Oncifelis geoffroyi*) resides in southern parts of South America. Its range is largely within Argentina, but also within southern Brazil, Paraguay, Uruguay and down into Patagonia (Berrie, 1978; Johnson and Franklin, 1991). It is a small cat, roughly equivalent in size to the domestic cat. The basic colour of its hide is a smoky grey, covered with numerous small black spots on the sides and limbs. The undersides are cream coloured. The tail is ringed with black bands, while the cheeks are characterized by two dark streaks and the neck with long dark lines. Its hide is in demand, both in South America and formerly in Germany. After the bobtail, the Geoffroy's cat's pelt is the second most heavily traded in the world. As an example, between 1979 and 1980, over a quarter of a million Geoffroy's cat skins were traded internationally.

The various habitats of this ground-dwelling cat include grasslands, marshes, shrublands, steppe lands and tree cover (Lucherini *et al.*, 2000). It has the habit of standing upright for lengthy spells on its hindlegs with the support of its tail and can climb trees with ease, often defecating in the forks of tree branches and reusing these places for defecation.

During the day the Geoffroy's cat typically rests. Like many other cats, its crepuscular nature makes it most active before dawn and after sunset. Its prey includes hares, small birds, rodents, frogs and fish. The hunting range for this cat may be as small as $2\,km^2$ or as wide as $10\,km^2$.

The Geoffroy's cat may breed during any time of year, but females usually have their oestrus cycles in autumn. Oestrus can vary from 1 to 10 days. Mating is performed in the usual feline manner with several males. Pregnant mothers are known to use their own fur to construct a birthing nest, pulling it from their coats. Litter sizes generally number one to three kittens, born with closed eyes until about 19 days of age. Their development is slow, and by 7 weeks they are eating solid food, at which age weaning begins. At about 6 months of age, the kittens are now fully grown and begin dispersing on their own. Although this cat is tolerant of habitat change from agricultural spread and deforestation, its greatest challenge to survival is excessive hunting. The European Economic Community banned the importation of Geoffroy's cat pelts in 1986.

Leopard Cat

Similar in size to the domestic cat, the leopard cat (*Prionailurus bengalensis*) is a notable Asian cat with an extensive distribution, inhabiting 21 Asian countries (Lim, 1999). It is widespread in China, the far east of Russia, the Phillipines, Pakistan, India and Malaysia, with a population in Sumatra (Santiapillai and Supraham, 1985). It has the greatest geographical extent of any small Asian cat. Notable features include a slim shape and variable coat colour that is generously spotted. It has an array of names in different regions as a result of the belief that its varieties were different species of cats. With an abundance of prey, namely rodents, this cat can become obese in appearance with excessive hunting (Rabinowitz, 1990).

The leopard cat has tall rounded ears and long legs, despite its small size. Other physical characteristics include a lengthy coat, webbed toes, a short narrow muzzle and a long tail. The weight may vary from 3 to 8 kg, depending on the region in which it resides. Its basic coloration may range from tawny or ochre to silver or brown, covered with black spots of differing sizes over the body, legs and tail. The head is small and has a pair of black stripes running up from the eyes to the crown. Another pair of black stripes is drawn across the cheeks, while two white line markings extend from the sides of the nose to the forehead.

The leopard cat is an able swimmer and frequently goes into water, sometimes to fish. It can live in a wide variety of habitats from tropical rainforests to dry coniferous forests, coastal areas and river valleys. Forest cover, however, is usually a preferred habitat.

This cat is active during the day or night, often hunting rats in trees. In addition to rodents, the leopard cat feeds on lizards, birds of various types, domestic poultry and fish. The average home range is about 2 or 3 km^2. They mark their territories with scrapes and faeces, sometimes covering their excrement. Both sexes spray urine during marking.

The female may breed any time of year, but will do so only once annually. Oestrus is short and mating is performed in the usual feline manner. Birth dens are chosen from any concealed place, such as amid bushes or the hollow of a fallen tree. Litters usually consist of two or three very small kittens which open their eyes at 10 to 15 days of age. They grow quickly and are able to eat solid food when they reach 1 month old. At 6 months of age they are full grown, with males larger than females. Captive leopard cats have lived for 13 years, and this is taken as their natural lifespan. Many have shorter lives as a result of being hunted for their pelts. The Chinese government, for example, has set a quota of 150,000 pelts per annum. Sadly, this cat is likely to disappear from parts of its extensive range as a result of such hunting.

Oncilla

The oncilla (*Leopardus tigrinus*) is an extremely small cat, one of several tiny cat species in South America. Its distribution covers most countries in the northern half of the continent, including Colombia, Venezuela, the Guyanas, Paraguay, Peru and Brazil. Its numbers are greatest in Brazil, as over-hunting has diminished numbers elsewhere. It has a variety of common names in different countries as it was not properly recognized until recently.

This animal weighs on average just 2 or 3 kg as an adult. It lives at varying altitudes and terrestrial conditions including cloud forests, subtropical forests, wet and dry savannahs and thorny scrublands. It has thick, soft fur with a background colour of a light red shade, covered with black spots that are generally arranged in longitudinal rows. Its spots include the head, neck and legs, while the tail is ringed. This small cat is lightly built and has a fine head with rounded ears (Sunquist and Sunquist, 2002).

The oncilla hunts any time of day or night and eats a wide variety of small creatures, including lizards, small rodents, grasshoppers, beetles and birds. Little is known about its general activities, however, as it is now rare in the wild. Information from zoo captivity indicates its oestrus period to last up to 9 days. The litter consists of just one kitten, with weaning occurring at about 12 weeks. The young are not full grown until about 1 year of age.

This small cat is very tolerant of people and has been known to wander into cities. Formerly, it was heavily exploited in the fur trade, but has now come under such strong legal protection that exportation from its native lands is prohibited.

Margay

Of intermediate size between the ocelot and the oncilla is the margay (*Leopardus wiedii*), another small cat in Central and South America that has been extensively hunted for its pelt. Apart from differing sizes, the margay and the ocelot closely resemble one another (Pocock, 1941). To distinguish between the two, the margay has very large eyes and a longer, bushy tail marked with a black tip and numerous black rings. The oncilla has a shorter tail and legs, and its spots are not open as displayed in the margay, which has pale-centred and open spots arranged in longitudinal rows. These two cats are genetically linked. They are also linked to the Geoffroy's cat, as the latter belongs to the ocelot family of cats in South America.

The distribution of the margay is extensive, covering all countries from Central America to Uruguay. This an arboreal cat, living in dense forest of various types, including cloud forest, dry tropical forest and humid tropical forest. It is particularly agile, even acrobatic, in its tree climbing. Special anatomical features such as a metacarpus that can rotate through 180° and broad feet with mobile toes allow for such acrobatic movement. These features allow the cat to grip tree branches firmly while moving throughout them with speed. The cat's capability for tree living is aided by its tail, which provides an effective balancing pole in its movements (Sunquist and Sunquist, 2002). The margay performs most of its hunting in trees, where it feeds on large mammals such as monkeys and birds. It has excellent eyesight for this manner of predation. Additional predation occurs at ground level, where common prey includes mice, rats, rabbits, agoutis, guinea pigs and opossums.

In its territory, this cat performs the usual marking practices of the majority of felines; scraping logs and branches, spraying urine and leaving faeces either covered or uncovered depending on the location. Margays live solitary lives and are quite rare in their native regions due to over-hunting and the loss of territory from deforestation. Until very recently, they comprised one of the four most heavily exploited spotted cats. During the latter part of the last century, export figures on margay pelts ranged from about 20,000 to 30,000 per annum. In spite of restrictive regulations on their exportation, there had previously existed a system of transferring pelts to Paraguay. From here, pelts were sent to Germany by the thousands. Recent international legislation has been effective in controlling the trading of margay pelts.

The margay's breeding habits are in line with the common feline system. Mating is in the usual method, with the female demonstrating oestrus, followed by multiple matings with more than one male. The courtships and matings are usually performed in trees. The litter usually contains just one kitten which opens its eyes at about 2 weeks of age. By 10 months the young animal is fully grown. Longevity under wild conditions is unknown, like much of this cat's biology. However, one such cat residing in captivity in Florida lived to be 24 years old.

Puma

As the Eurasian lynx is the most widely distributed cat on the European and Asian continents, the puma (*Puma concolor*) is as widely spread on the American continent (Stocek, 1995; Spreadbury *et al.*, 1996; Nuñez *et al.*, 2000). Its territory spreads from Canada in the northern hemisphere to Chile in South America. Puma is the original native name for this cat. In North America it is the largest wild cat, and is commonly called the 'cougar' or 'mountain lion' and occasionally, the 'panther', while the true name 'puma' prevails in most of South America.

The numbers of pumas in the southern USA and in Central and South America have been greatly reduced due to hunting and loss of natural territory. It is still well established, however, in mountainous areas, tropical rainforests, scrublands and seasonally flooded lands. The latter include the Pantanal in Brazil and the Llanos that spreads from eastern Colombia across into Venezuela. In these latter areas it avoids conflict with the jaguar by limiting its territories to parts unoccupied by this larger cat. In the Pantanal, pumas inhabit the drier regions, while jaguars prefer wetter habitats. In the Llanos, the pumas occupy more open and drier regions, while jaguars prefer the cover of jungle. The wide variety of terrain utilized by the puma illustrates its great adaptability to varying habitats (Johnson and Franklin, 1984; Iriarte *et al.*, 1991). The puma is well established in Central America and in Costa Rica it is proudly portrayed on the national emblem.

The puma's weight varies from region to region. In its most northern and southern ranges it is about twice the size of its species occupying the tropics. Adult males in equatorial forests weigh as little as 30 kg, while Canadian puma males can weigh up to 80 kg, with females weighing less in both areas. In either weight range, the puma has remarkably long hindlegs, giving it spectacular jumping ability. A plain-coloured cat, its coat varies in shades of brown, with a whitish muzzle and undersides. The ears are rounded, while the lengthy, cylindrical tail has a black tip (Fig. 8.1).

The puma lives on the ground, but when necessary can swim and climb trees very efficiently. It typically rests during the day and hunts at dawn

Fig. 8.1. Puma head; note large nose and small eyes typical of large cats.

and dusk. Many pumas travel extensively while hunting – up to 30 km in one night. Like most felines the puma is an opportunistic predator, preying upon that which it encounters. As a result, it is known to eat a wide variety of catches, from small rodents to young moose. In North America the principal prey is deer, but raccoons, pigs and armadillos are also sought after. In South America the puma also consumes deer and additional mammals of smaller size, including peccary, capybara, guanacos and even young alligators. In locations with domestic stock, pumas are a major threat to calves, sheep, cattle and horses (Hayes *et al.*, 2000).

In the customary feline fashion, pumas hunt by creeping close to their prey before rushing and pouncing. The distance rushed is seldom more than a few metres. Prey killed in the open is always dragged away to a concealed location prior to consumption. When portions of the carcass are left over, they are concealed with grass, twigs and soil for devouring later, either singularly or with cubs. Females accompanied by cubs kill much more frequently than those without young. In maintaining its home range, the puma leaves its mark by scraping patches of the ground and building mounds of faeces covered with soil. Unlike its relatives, it does not spray urine. Both males and females cover their faeces. Occasionally, the puma emits a piercing scream to announce its presence (Smallwood, 1993).

During their breeding season female pumas exhibit oestrus in the common cat fashion – rolling and calling. At the onset of oestrus the female's behaviour is a mixture of aggression and solicitation, as is another common feline display prior to mating. Oestrus may last about a week and males consort continuously with the female cat during that time. During copulation, both male and female make screaming sounds. These sounds are equivalent to the caterwauling that is such a feature of domestic cat breeding.

The pregnant puma creates a birthing den in a well-concealed place such as a cave, a pile of rocks, a thicket of foliage or a tangle of uprooted stumps. The litter size is usually two or three kittens and, at first, the young have very thick, brown spotted fur. This neonatal coat gradually changes with age, and has grown out before the young animal is 1 year old. The kitten's eyes open at 2 weeks, followed by abundant activity in their play and general movement. Kittens are not weaned until they are more than 2 months old, at which point they accompany their mother on hunting excursions. Once they reach 1 year of age their independence has now developed, followed by dispersal as they seek new territories. The lifespan of the puma is variable, but they are known to live into their teens.

Sand Cat

The sand cat (*Felis margarita*) is a very specialized species, living where few others could flourish. It is found in areas of sandy desert, including Algeria, the Middle East, Pakistan, the Arabian Peninsula, Israel and Turkmenistan by the Caspian Sea. Well adapted to such habitats, it lives in burrows and is the only cat whose distribution is confined to desert regions.

A very small cat with a stocky build, it averages between 2 and 3 kg in weight. It has a long tail and short legs that appear to create a wide stance. Its overall colour is suitably sandy, with dark bands on the legs and tail. The tail and ears are black tipped, while a dark streak runs from the outer corner of each eye across the cheek; its underside is white.

In areas of its territory that are cold at night, the sand cat has a dense coat. A unique feature is the presence of long, thick hair growing between its toes, to the extent that even the pads are covered, insulating and protecting the soles of the feet as it travels on hot sand. The sand cat has a special manner of movement. It can run for distances of about 400 m at speeds about 40 km/h, keeping its belly close to the ground as it darts with periodic leaps. Similarly when hiding, it lies flat on its belly with its chin on the ground and its huge ears pointed down.

The temperatures in the desert sands are extreme, being very cold at night and intensely hot during the day. The sand cat adjusts to these temperatures by retreating into its burrow during the extreme heat or cold. While this cat will use burrows abandoned by foxes or porcupines, it generally digs its own burrow in compact sand or soil at the base of a shrub or bush; its burrow slopes down for about 2 or 3 m to a depth of about 60 cm. Its custom is to rest at the mouth of the burrow when temperatures are moderate. When bad weather strikes, however, it can remain here for several days.

During its nocturnal hunting excursions, the sand cat feeds on small mammals such as gerbils, jerboas, voles and hamsters. In the day it hunts lizards and snakes, and will also catch birds when given the opportunity. Sands cats have a reputation in the Sahara and Israel as snake hunters, killing and feeding on vipers (Sunquist and Sunquist, 2002).

The domestic cat has been described in this text as a water miser (as are most wild cats), but none can match the sand cat in water conservation. The latter survives without drinking water for extended periods, and does not drink at all for much of the year. Evidently, it acquires its necessary amount of fluid from its prey. Within captivity, with water readily available, one individual has been observed to not drink for 2 months. If water is available, however, sand cats will usually drink regularly.

The sand cat's method of hunting is active searching rather than lying in wait. For its size, it travels great distances while hunting – often 6 to 10 km per night. Within its territory, it scent marks by rubbing its cheeks on prominent objects and leaving claw marks wherever possible. It also sprays urine, but covers its faeces.

This cat can breed during any time of year. The duration of oestrus is about 6 days. Mating is done in the ritualized feline fashion, with the male biting the female's neck as he crouches over her during repeated copulations. The female commonly has two litters per year. Litters are relatively large in number, ranging from three to eight kittens. The kittens are fully grown by 1 year of age. There is no information currently known about their dispersal or longevity.

The sand cat is not exploited for its coat, but the ability to catch this animal with ease led to its capture for zoos, and its numbers declined in such regions as Pakistan. This feline, however, does not cope well in captivity and a high mortality rate has caused its stoppage in the zoo trade. Its high reproductive ability allows this cat to survive the natural population fluctuations that occur. It appears to have no fear of human approach and easily tolerates being kept as a domestic pet.

Jungle Cat

The jungle cat (*Felis chaus*) in north-east Africa is widely understood to be the progenitor of the domestic cat, although it is larger than the housecat and has a lankier appearance. In spite of its common name, the jungle cat does not inhabit any jungle. Its geographical distribution is very extensive, ranging from the Middle East to south-west China, including the subcontinent of India, where it is considered a major species of wild cat. Within its great distribution, the jungle cat commonly has its habitat at lower altitudes close to wet areas such as swamps, extensive marshes, floodplains, streams, reservoirs, lakes, ponds and densely vegetated river valleys. Banks of rivers are favoured, giving the cat a riparian affinity. This cat is also capable of living in dry regions within Nepal and India, for example. Its variety of habitats demonstrates the jungle cat's environmental adaptability. This, in turn, explains its distribution in Asia, in which it is the most widespread wild cat.

The jungle cat creates its home in dense cover, including thickets, crevices and burrows. It has been observed sunning itself near its hiding place. Such behaviour is also witnessed in the sand cat (and similar to the housecat's habit of seeking sunshine through a window). The jungle cat hunts at night and often travels up to 5 km in search of food. Occasionally, it moves about during the early and later parts of the day. It maintains a solitary life except at breeding time.

Small rodents such as rats, mice and gerbils constitute its principal prey, while squirrels, frogs and birds are also consumed. The jungle cat is physically able to run at speed and jump in a high, arching leap in its hunting practices. Additional prey includes, whenever possible, lizards, fish and snakes. If it dwells in close proximity to human settlements the cat will readily feed on available poultry. This great variety of acceptable food allows the jungle cat to cope well in an array of circumstances.

When breeding, the jungle cat behaves exactly like the domestic cat with the usual caterwauling and tentative approaches, followed by the typical feline mating routine. The male crouches over the female, biting the nape of her neck. Following copulation, the female rounds quickly on the male with a cry and attempts to bite him aggressively. For the birth, the female makes a birthing den for the litter, which typically contains three kittens, although larger litters of up to six can occur. The kittens open their eyes at about 12 days of age and continue to suckle until they are about 3 months old. By 9 months of age the young cats are mature and take to living on their own. Their natural lifespan appears to approach 10 years.

As to the jungle cat's well-being in the near future, it seems to cope well with losses of territory through its adaptability and its wide variety of acceptable food. Although it formerly suffered from heavy killing for its pelt, the Indian government has now taken action and has banned the export of these prized coats. The killing rate a few decades prior to this numbered roughly 300,000 cats per annum.

European Wildcat

The European wildcat (*Felis silvestris*) has a vast continental distribution, dwelling in forest regions in Europe (Stahl, 1993). It is found in France, Spain, Portugal, Belgium, Germany, Italy, Poland, Switzerland, the Czech Republic, Slovakia, Romania, Hungary, Bulgaria, the former Yugoslavia, Greece, Turkey, Georgia, Azerbaijan, Armenia and the Carpathian Mountains of Russia. It also inhabits the islands of Sicily, Sardinia and Cyprus. The wide distribution of this species indicates its very ancient history.

The continental European wildcat is compact in build with short legs. It is grey brown in colour with a light grey underside and dark stripes over its head, limbs and back. Black individuals have been reported, but none on the European continent.

The European wildcat preys upon various birds, young deer, water voles, muskrats, martens, polecats, young chamois and a variety of small prey including insects, as well as grasses and fruit. The wildcat in Europe has an ever-increasing territorial problem from human encroachment into its remaining habitat, but efforts are being made for some protection for this cat (Nowell and Jackson, 1996).

African-Asian Wildcat

The African-Asian wildcat (*F. silvestris librica*/*F. silvestris ornata*) is distinctly different from the European wildcat in its appearance, namely colouring and physical build. One visible differential feature is the colouring of the back of the ears. A brick-red or buff shade characterizes those of the African-Asian wildcat. Further, this cat has much longer legs, such that when it sits on its hindquarters, the forelegs hold the cat in a very upright, almost vertical posture. These long legs create a cheetah-like stride. It does not normally exhibit the tabby cat appearance of the European wildcat as its colouring is varied, mainly greyish or reddish with spots that form transverse bars, particularly across the legs and around the black tipped tail. The tips of the ears have a short but pronounced black tuft of hair. The undersides of the paws are black. In dry regions the overall coat colour is pale, while in humid areas it tends to be darker and more heavily spotted.

Like the European wildcat, this cat displays an extremely wide geographical distribution, spread over much of Africa and mid-Asia. In Africa, it is absent from the Sahara and the Congo where climates are dry, but common in southern parts of the continent. It is also found on the eastern border of the Mediterranean, around the coasts of the Arabian Peninsula, Iraq, Iran, Afghanistan, Pakistan and India's north-west plains. Its wide distribution illustrates this cat's adaptation to a variety of habitats in its search for cover and shelter for hunting and resting, respectively. Forestland, scrubland and grassland serve its purpose, and it can live agreeably with other small carnivores in such territories.

The African-Asian wildcat is strictly nocturnal in its hunting. Rodents constitute the primary prey. It is customarily a solitary hunter and an adaptable predator, catching an assortment of food when necessary or the opportunity strikes. Such alternatives include hares, rabbits, young antelope, birds, poultry, lizards, snakes, scorpions, beetles and even fruit. During a long-term drought in Botswana, this cat was found to have thrived on a diet of invertebrates, fruit and birds.

This cat is highly territorial, but has relatively small ranges varying from about 1 to 12 km^2. Both males and females spray urine and bury their faeces within their territories. Individual females defend their territories quite aggressively, perhaps due to the small size of their expanse. Male territories often overlap those of females.

African wildcats may breed during any time of year with a peak in breeding during a wet season. Oestrus lasts from 2 to 8 days and the female may come into oestrus repeatedly until pregnant. Two litters per year are common, with each litter size numbering about three kittens. The kittens are usually born in an underground den, a rock crevice or beneath a bush. They open their eyes at 9 to 11 days, nurse for about a month and then became active soon thereafter. At 3 months of age they begin accompanying their mother on her hunting trips and become independent at 6 months of age. Once mature, the kittens tend to remain in their birth area for a year before they begin their dispersal.

The lifespan of the African-Asian wildcat can reach the mid-teens under very favourable conditions. This species has actually had improved well-being as a result of agricultural development in many areas. Farming facilitates rodent populations, and this cat is inclined to move closer to areas where crops such as corn are being grown, in order to feed on the increased rodent population that associates with this.

Conclusion

These diverse species of wild cats show that felines inhabit the wide world, serving the natural purpose of maintaining balanced animal populations, including many that are not of advantage to mankind. It is unfortunate that the populations of these cats are diminishing everywhere, to the point that there appears to be an underlying feline genocide. Much of the loss of these species is now due to poaching and encroachment on habitat. The control of this can only be done by education, for which national regulatory bodies now have responsibility.

9 Minor Wild Cat Species

It is something of an arbitrary measure to separate the world's wild cat species into a principal or minor category, but the fact is there are species with small populations or limited distribution. This fact designates the need to group them separately. In addition, some of these wild cat species are atypical felines. As an example, three such cats discussed in this chapter are particularly small, weighing roughly 2 kg, with one such species residing in South America, one in Africa and the other in Asia. Their unique features include a cat that can swim under water, another which resembles a weasel and a third species whose head resembles a Pekingese dog. Nevertheless, all are members of the cat family with a heritage that spans more than 10 million years and all are deserving of human awareness and freedom from human abuse.

Bay Cat

The bay cat is the rarest cat in the world and the species about which least is known. This cat (*Catopuma badia*) is found only on the island of Borneo. As a separate species, its history stems back for roughly 5 million years, before Borneo even became an island. The bay cat of Borneo is approximately the size of a large domestic cat, but has an exceptionally long tail. Its coat is characterized by a fairly uniform, basic colour varying from mahogany red to dark grey, with faint spots on its undersides and legs. Faint lines are visible on top of the head and cheeks, while the hair about the head and throat grows forward. The ears are short and rounded with a low placement on the head. This species resides in dense forest, but little else is known about its habits. The rarity of this cat may make it heavily sought after by zoos. Its value is already estimated at US$10,000, and while it is protected against trading and hunting, the occurrence of illegal capture could subject the cat's well-being to great risk on its native island.

Black-footed Cat

The smallest wild cat in Africa is the black-footed cat (*Felis negripes*). Like some other cats, only the soles of the feet are black, and it could be said that this species is misnamed. Its weight measures in at roughly 2 kg, categorizing it as one of the world's smallest cats. The kodkod of South America and the rusty-spotted cat of India are of the same diminutive size and weight, awarding these three species the inclusion in the world's smallest cat category. The black-footed cat is only found in southern Africa in the countries of Botswana, Namibia and the Republic of South Africa. Like most cats, its eyes are large in relation to its head size. The ears are also large and rounded. The tail is short, with a black tip. The coat's background coloration is an orange brown, coated with conspicuous dark spots that merge with each other to form bands on the neck and legs, while creating rings around the tail.

The black-footed cat lives in dry and waterless areas with partial vegetation cover. It can live on the fluids of its prey for long periods if water is not available, but can also drink on a normal schedule when water is abundant. Its activities are strictly nocturnal, and it spends the daytime resting and hiding in dense cover, burrows or empty ant hills. Active at night, it may travel considerable distances before returning to its home base before dawn.

In spite of its small size, the black-footed cat is an efficient hunter. Its prey includes gerbils, mice, shrews, small birds, locusts, grasshoppers, moths and, at times, young springbok. Occasionally, this feline will sustain itself with a large kill such as a hare, but depends primarily on rodents, which can be killed in large numbers with great capability. Their hunting tactic will often involve waiting cautiously outside a rodent hole for long periods, listening for any sound. Despite its size, the black-footed cat can run quickly and pounce 2 m when catching prey.

In their social organization, black-footed cats have solitary lives and follow the common feline arrangement. A male has a territory which usually overlaps those of females, but does not overlap another male's. The male territory spans between 10 and 15 km^2, while female territories are often about 5 km^2. Both sexes mark by spraying, scratching and leaving their faeces uncovered. They are also capable of emitting an extremely loud call, similar to the roar of a tiger but in a higher pitch.

Black-footed cats engage in both friendly and aggressive encounters amongst themselves. During displays of aggression, particularly towards humans, they tend to spit, hiss and growl impressively, while holding their ears firmly flattened to the side.

The female's oestrus period lasts just a day or two, and her duration of male acceptance is only about 10 h. During this time, numerous copulations will occur. Litters are born in natural hollows in the ground. A female will often have two kittens, which are active from birth. Their eyes remain closed until they are between 2 and 10 days of age. The mother changes their location every week or so, carrying her kittens by the scruff of the neck to each new den – a habit thought to be a tactic in avoiding predators such as hyenas and jackals.

The kittens can walk at 2 weeks of age and start to engage in playful activity. At 6 weeks they can run from the birthing nest, and by 1 month old they can eat solid food. Their rapid development renders them invulnerable to predation. This species is extremely rare. Its well-being is assured by its size and secretive manner of living in daytime, making it an obscure animal in its arid world.

Chinese Desert Cat

Due to its extreme rarity, the Chinese desert cat (*Felis bieti*) has not been studied sufficiently enough to have its classification verified as a species or subspecies of its own. It is roughly twice the size of the domestic cat, with a thick grey coat and black outer hairs. The underside is a lighter shade, while faint stripes run across the legs. The backs of the ears are black, while tufts of this hair extend from the tips. The end of the tail is also black.

The range of this cat is limited to northern China, where its various habitats include mountains, meadows and semi-desert. It preys on rodents chiefly, but will also hunt birds. Its hunting time is crepuscular and it spends daytime resting in holes in the ground. The female will give birth in burrows, and litters commonly include two kittens that develop in the usual schedule of epigenesis in mid-sized cats. With such remoteness from human activity, their collective well-being is not likely to change.

Asiatic Golden Cat

The Asiatic golden cat (*Catopuma temminckii*) resembles the African golden cat but is slightly larger with a longer tail. The two cats are clearly related, but have been classified as separate species. Both are described as cats of many colours. The Asiatic's coat can be uniformly golden brown, red, grey or brown. White lines run from the inner corners of the eyes up to the crown of the head and across each cheek.

The distribution of this species extends from Tibet, southwards to Malaysia and the island of Sumatra. Its usual habitat is in forested areas at different altitudes, but open rocky areas are also considered. It marks its territory in the typical feline fashion by urine spraying and scratching. The rate of urine spraying in the female increases during oestrus, as it commonly does in many other felines. A night hunter, the Asiatic golden cat preys upon rodents, birds, lizards and even small deer. It is also capable of killing small domestic stock such as sheep, goats and calves.

The female's oestrus period lasts about 6 days and can occur at any time of the year – another common feline characteristic. At copulation, the usual nape bite by the male is brief. Pregnant females often give birth to only one kitten. The kitten opens its eyes variously between 6 and 12 days of age. It grows rapidly and can walk by 2 weeks. Sexual maturity is attained by 2 years of age and its lifespan extends about 15 years.

Regarding their communal well-being, the advance of the timber industry throughout their region has deprived many Asiatic golden cats of their habitats. Since this species is chiefly forest-dwelling, and does not breed well in captivity, its long-term prospects are poor.

Manul

The manul (*Otocolobus manul*) is an atypical cat that inhabits central Asia. While similar in size to the domestic cat, its long, thick coat makes it appear much larger. Its body weight is only between

2 and 4 kg, and its conformation is squat and short-legged with a long, bushy tail. The broad, low forehead and flattened face create a striking resemblance to a Pekingese dog. The widely placed, furry ears are positioned on the head in a low set fashion – as such that the tops of the ears are at the level of the eyes.

The appearance of the manul's coat is pale and shaggy, with a grizzly shade of rust, silver or ginger. The tail has a black tip and black rings. White rims around the eyes, lined with black, create the impression of spectacles.

The manul has a wide distribution in the central Asian uplands and Arjun Mountains of western China, residing at various altitudes. Its habitats include steppe lands with vegetation or rocky hills and valleys. In their home territories, they spend a portion of the day in a den, or sunbathing if the weather is suitable. They inhabit their dens all year round to provide themselves with protection from the hot and cold extremes of their continental climate. The bedding of a den consists of feathers, dried vegetation and pieces of skin from prey. Unlike most cats, the manul hunts during day or night, feeding on partridge, voles, gerbils and other rodents.

The manul is a seasonal breeder, and litters are born in April and May. The female's oestrus lasts 1 or 2 days. The mating process includes vocalizations by both sexes and several males participating in copulation. Litters are usually large, with kittens numbering four to six in an annual litter. The young are born with dense, fuzzy fur and following roughly 2 months of age they moult, acquiring the adult coat. By 5 months, they are able to hunt for themselves. Their lifespan is about 10 years. Captive manuls are quite friendly, and generally show no fear of humans. Their breeding success in captivity, however, is poor.

The manul has a fairly steady population in western China, and is listed as a protected species in this region. Its survival prospects are encouraging.

Flat-headed Cat

Another atypical feline is the flat-headed cat. The distribution of this species is limited to the Malaysian region of the Malaysian Peninsula and the islands of Sumatra and Borneo. Its physical appearance resembles an otter, as opposed to its feline family. Its claws are only partially retractable. The eyes are closer together and further forward than in the usual facial eye-set. The flat-headed cat is elusive and very rare in its geographical region.

Unsurprisingly, this species has an elongated, flattened skull with small, rounded ears. The elongated head permits the animal parallel rows of pointed cheek teeth. Its fur is thick with a fine texture, characterized by a brown roan colour over its body and a reddish brown head. The face has a lighter colour and the throat and muzzle are white.

The habitat of the flat-headed cat is always close to a waterway in primary forest, such as a well-treed riverbank. Essentially a fishing cat, it lives a solitary life close to water, feeding variously on fish and rodents. Nothing is known about reproduction in this rare species. Although currently under the threat of lost territory, there are signs that it can adapt to life in oil palm plantations. These plantations are replacing the primary forest that constitutes its natural habitat. Such adaptation is the only way this animal can have some future prospect of collective well-being.

Fishing Cat

Like the flat-headed cat, the fishing cat (*Prionailurus viverrinus*) preys on fish, which comprise a large proportion of its diet. While it is not alone in this habit, its title gives it a distinction. A large cat, it is roughly twice the size of its domestic cousins and is covered with stripes and spots. The undersides of the body are white, and the head is round with black ears. It has a powerful appearance, heightened through a deep chest. Its legs are short and the toes on the front feet are webbed. The black-tipped tail, while short, is strong and thick at its base. The tail is believed to function as a rudder when the animal is swimming. It can swim for extended periods, descending beneath the water's surface as it hunts for fish. The fishing cat can live in groups.

As could be expected, this cat's habitat is always in close proximity to water, concealed in thick cover. Its territory includes various Asian locations, extending from Pakistan, Nepal, India, Sri Lanka and Thailand, to Vietnam and Malaysia. A male's territory is about 20 km^2, while a female's is about 5 km^2. The male territory generally overlaps several female territories. Territorial marking is performed by spraying urine and rubbing its head on to environmental structures. Calling loudly with a gurgling sound is also used to indicate possession of the home range.

The fishing cat is an aggressive hunter, with a history that includes killing calves, dogs and even young children. Its teeth are well-suited to killing mammals. A seasonal breeder, the fishing cat will usually mate in January and February. Litters of two or three kittens are then born in a den. The kittens can walk and climb when they are 1 month old. At 2 months of age they become quite active outside their den, but continue to suckle for long periods, such as 6 months. At 1.5 years of age, they are sexually mature. In its old ranges, the fishing cat has become rare due to a combination of habitat loss, hunting and being destroyed for its predation on small domestic stock. It is commonly found, however, in plantations, to which it has made some adaptations.

The disposition of the fishing cat tends to be wary and aggressive, yet they have been captured and successfully kept as pets. A future in captivity, however, is bleak. Similarly, its existence in the wild state appears to be in doubt, in the long term at least.

Rusty-spotted Cat

The rusty-spotted cat (*Prionailurus rubiginosus*) of India is the Asian member of the world's tiny cat trio. It weighs about 2 kg, making it less than half the size of the average domestic cat. It resides in southern India, Sri Lanka, Kashmir, and in a portion of the Gir Wildlife Sanctuary and National Park. Within its range, it inhabits deciduous, tropical and scrub forest, bamboo growth, shrublands, grasslands and rocky areas at various altitudes.

This tiny cat is a very energetic animal, displaying bursts of activity during the day, but most active at night. Being boreal, it is capable of running gracefully up a tree in a squirrel-like fashion. Like many other cats, it sunbathes or reclines in cover during the day. On the ground, it moves with quick darting movements at a faster rate than other cats of its size. This practice, combined with its fierce reputation and rapid tree-climbing ability, provides protection from larger predators, such as foxes, jackals or larger cats.

The coat is short and reddish brown in colour, while the head is marked with several black stripes. Rusty brown blotches are arranged in a pattern of lines on the back and flanks. Although the undersides of the body are white, there are traces of rusty brown spots. The ears are small and rounded, while large eyes relative to its body size create a kitten-like appearance. This latter feature is common in many cats within the smaller-sized categories. The rusty-spotted cat hunts mainly on the ground and feeds on small mammals, birds, lizards, frogs and insects. If given the opportunity, it will also eat poultry. In captivity, it is reputed to be a ravenous eater.

This species does not appear to have a distinct breeding season. The duration of oestrus is about 5 days and the process of copulation includes the usual feline rituals, including vocalizations, straddling, frequent matings involving several males, male nape biting and post-coital attacks by the female. The pregnant female gives birth to just one or two kittens in a den. The rusty-spotted coat does not present itself on the young right away. Their lifespan can extend to 10 years or more.

Rusty-spotted cats have been adopted as pets and accept a domestic existence with ease, but do not always retain their well-being in captivity. They are highly susceptible to infection with the feline enteritis virus, causing a high mortality rate. Any attempt to keep this cat captive would require prompt vaccination against the disease. As a result of its tendency to live near human habitation, this cat is subject to human disapproval and hunting and has become rare. Its survival prospects are in question.

Kodkod

The kodkod (*Oncifelis guigna*) of Chile could very well be the world's smallest cat. Weighing about 2 kg on average, it is much smaller than the average domestic cat. It has a dull-brown or buff-coloured coat, covered with small brown spots. The tail is short and bushy with black rings, while large paws dominate its size. A few specimens may be completely black (melanistic).

This tiny cat makes its home in the southern part of Chile and may be active either day or night. It is a remarkable traveller for its size, often covering 5 km during its daily excursions, which may include tree climbing through forests. The male occupies territories of about 2 km^2, which includes the home ranges of several females. Predation is mainly on rodents and birds, but small domestic livestock, including poultry, are also targeted.

Destruction of natural habitat is the greatest risk to the existence of this rare species. Fortunately, legislation has been introduced in Chile and Argentina to prohibit the sale of this cat and its pelt, providing as much protection as is realistic. Its future prospects, therefore, appear to be fairly positive.

Andean Mountain Cat

The Andean mountain cat (*Oreailurus jacobitus*) is an immediate neighbour to the kodkod and, like this latter species, it is very rare. Few observations have been made about the Andean mountain cat. Its size is comparable to the housecat. Its coat is characterized by long grey fur and striped with partial dark markings. The white underside has black spots and the tail is long and bushy with several black rings. The legs have similar bands of black.

The geographical home of this cat is a section of the high Andes in southern Peru, Bolivia and northern Chile. This area contains the high plateau of Altiplano, with an altitude of 4500 m where daily temperatures usually fall below 0°C. In spite of such freezing temperatures, these cats inhabit such areas where there are various grasses, plants, bushes and shrubs. Such vegetation sustains the prey of the Andean cat, which includes rodents, rabbits, armadillos, viscachas, ducks, geese and other birds.

Although it is now illegal to hunt this species, the killing of this cat continues to prevail. As a result of such poaching, combined with its rarity, the Andean mountain cat is likely to become extinct in the near future. Its protection thus far has been granted through its inaccessible location but, sadly, determined hunters who know of its whereabouts may end that security.

Jaguarundi

Like the manul and the flat-headed cat, the jaguarundi (*Herpailurus yaguarondi*) is another atypical feline in many of its physical and behavioural features. A long body and tail supported by short legs give it the appearance of an otter or weasel, particularly when it is scurrying in undergrowth. Its coat is plain, and may be greyish or reddish, with some variations in shade between its two colour phases. A litter may contain kittens of either colour, and it is not uncommon for some individuals to have black coloration. The average weight among adults is 4 or 5 kg.

The jaguarundi has a wide distribution, covering virtually all of South America and Central America at various altitudes. The territories that it occupies are often in open areas that can provide partial cover with bushes or small trees. They readily travel and hunt during the day, and are characteristically more diurnal than nocturnal. Their primary food is rodents, but as opportunistic predators, they will also feed on rabbits, lizards, armadillos, guinea pigs, birds and any available fish. Insects, fruit and grasses may also be consumed.

The jaguarundi will often go in pairs as a permanent unit; socially, they can form occasional groups. Shared territory is common, with ranges measuring anywhere from 8 to 100 km^2. They journey through their territories quite actively, marking them with uncovered faeces and urinating (but not spraying) on their scratch marks. At certain sites they may rub their heads against objects.

In tropical regions, there is no occurrence of specific breeding seasons. The duration of oestrus is 3 to 5 days and the usual feline mating rituals are followed. Born in a den, the litter size may range from one to four kittens, with furry coats at birth. At 6 weeks old, weaning commences and the mother supplies her young with prey. Since this species is not targeted for its pelt and has the capacity to exist in a wide variety of locations, climates and habitats, its survival prospects are good.

Marbled Cat

The marbled cat (*Pardofelis marmorata*) has a similar geographical distribution to the fishing cat and the flat-headed cat. It is about the size of the average housecat and has a long bushy tail. Its hide has a woolly base and its coat pattern resembles that of the clouded leopard – a greyish undercoat covered with irregular dark blotches and black spots involving the head, neck and ventral areas.

Its habits are of a nocturnal and arboreal nature, living and hunting in trees in remote areas. At night, this species displays much activity, including jumping and climbing. It preys chiefly on birds, but lizards and rodents are also sought after. Litters are small, and kittens are walking at about 10 days of age once their eyes have opened.

Unlike some of its aforementioned relatives, the marbled cat is particularly reclusive and does not accept proximity to human habitation. It vacates its territory promptly when any sign of human activity is present. The rarity of this species is facilitated continually through the lumber industry, and it is likely to cause its extinction in due time.

Pampas Cat

Although it appears to share the size of an average domestic cat, the pampas cat (*Oncifelis colocolo*) is actually smaller, averaging about 2 to 3 kg in weight – its long hair creates the illusion of a larger size. The fur may be soft or coarse and it forms a mane in some individuals. The colour pattern is varied. The background coloration may be whitish, greyish or silvered, with reddish streaks or spots on the sides; this overlayer is often without patterns. The legs, however, are distinctively marked with brown bands, while the tail has brown rings.

The pampas cat is South American and its distribution extends from Peru in the north to the pampas of Patagonia in the extreme south. Within this range it occupies a great variety of habitats, namely grasslands, but it is also found in the mountainous Andean areas and in the great Brazilian Pantanal region that floods seasonally. This is an ecological contrast to the dry desert areas of Patagonia where it also resides. It hunts during the day or night, in trees or at ground level. It will prey upon whatever it can catch, including rodents, birds, domestic poultry and eggs.

Its reproduction has no seasonality and its manner of breeding is in the typical feline style. The female gives birth to one to three kittens in a sheltered nest. Little is known about this cat's other biological features. Although its pelt is not considered to have real commercial value, it was hunted until recently. Approximately 80,000 pampas cat pelts were exported per annum from Argentina in the 1970s. This trade has now been prohibited, but some population damage has already been done. In some countries it is also hunted for sport. The main challenge to its collective well-being is the loss of territory by the habitat destruction resulting from expansions in the agricultural and timber industries.

Scottish Wildcat

The Scottish wildcat has been known in Scotland from early historic times. This species is a branch of the European wildcat (*Felis silvestris silvestris*) living in the northern, highland and inner island areas. It resembles a large domestic cat, but has a broader head and thicker forelegs. Its ears are well spaced and its overall appearance is rugged, characterized by a long, shaggy coat. Its weight is generally about 6 kg, while some larger individuals have been reported.

The coat is a striped tabby on a grey background. Several dark lines extend from the sides of its eyes and over the brow. Additionally, a pair of dark lines runs across the cheeks from the outer corners of the eyes, while tiger-like lines cover the neck, limbs and back. It has long grey whiskers and a white muzzle and throat, while the undersides of the body are dark grey. The tail is thick with several black rings and a black tip. A few individuals displaying a black coat have been reported.

The common habitat of this species has plenty of cover and may be heather, forest or bushy scrub. It lives on the ground and occupies a semi-permanent den, hunting during the day or night. Its principal prey is the rabbit, but it will also hunt rodents, birds and weasels. Occasionally, if its territory falls in close range to remote human habitation, the Scottish wildcat will prey upon domestic poultry. Males and females acquire wide ranges, often about 180 ha where they enjoy solitary lives, except during the breeding season in February and March. During this time, both sexes travel in search of mates. At times they interbreed with ordinary domestic cats and this dilutes the wildcat's population (Hubbard *et al.*, 1992).

Within their territories, Scottish wildcats will leave their faeces uncovered on prominent places such as rocks, tree stumps and at the intersections of tracks. These faecal deposits are not randomly distributed, but are usually close to the animal's resting site rather than the boundaries of its territory.

Females have polyoestrus cycles until pregnant. Their oestrus period varies from 2 to 8 days, dependent on whether mating has occurred. The female may have two litters per year, often containing four or five kittens. The birthing place is usually an underground den or nest covered with dry grass or ferns. The young open their eyes between 7 and 12 days, and by 1 month are quite mobile. By 3 months they are weaned, and will then accompany their mother during hunting, sharing in the kill. Their lifespan has been reported to extend into the teens.

With the loss of territory due to agricultural development and reductions in rabbit numbers from myxomatosis, the Scottish wildcat's numbers have dropped dramatically in recent years. It is estimated that the total population is now less than 400, and unless measures are taken to provide further protection, extinction could occur within a decade. However, The Scottish Wildcat Association is attempting to maintain this cat's existence.

10 Common and Comparative Feline Features

Common Behavioural Characteristics

Self-living

Within their diverse populations, it is notable that solitary living is the feline norm. Although the domestic cat, when feral, has a predilection to form a social colony, it can live a satisfactory life alone and usually resents an intruder. The lone cat is able to form a partnership, however, with a compatible associate cat if the partnership is performed properly (Fig. 10.1).

Ecological solitude is the essence of feline nature. This fact makes each cat aware of its responsibility for its own maintenance. Top priority in feline dynamics is its own somatic needs. Self-interest ultimately prevails in an isolated existence. This self-based method of living contrasts with the social systems of herding animals. Their social singularity constitutes a major difference between cats and most other mammals. Felinity is hard-wired for enduring solitude, where all principal motivations relate to self-interest. Their genome shapes them as essential egoists, directing much of their behavioural characteristics and personalities. It is a fact that egoism is prevalent in cats, causing a compulsion to be free of importuning by competitive cats. At the same time, the feline ego is vulnerable to frustrations regarding the security of its status as a territorial holder.

When the compulsion of ego defence is aggravated in housecats, behavioural anomalies can develop. These anomalies are commonly in the form of spraying, soiling or aggression. In previously normal cats, these three forms of conduct are the most common types of abnormal feline behaviour. They are, in fact, manifestations of feline ego disturbance at a clinical level. While various treatments for these conditions have degrees of success, including surgery to olfactory receivers and pharmacological administrations with artificial facial hormones, it is likely that a natural approach could prove beneficial. Recoveries have resulted from social rehabilitation, involving several weeks of solitude with proper care in a cat-free environment. Unfortunately, the animal has to be returned to an environment that is free of feline competition.

Travelling

Most cats usually travel nocturnally. Lions travel in daylight (Fig. 10.2), thoroughly covering their territories.

A territory may be defined as the maximum area that the individual cat can patrol regularly. Males with greater strength are capable of travelling further than females. In exceptional circumstances of a radical reduction in prey, the predator is forced to travel much more widely in search of food. The Siberian tiger, for example, occasionally travels several hundreds of kilometres out its normal range in this search, as a result of a lack of natural prey in its own territory. The distance travelled regularly occupies the feline's entire night. Depending on the individual species, this distance can be a few kilometres for a miniature cat, to more than a hundred kilometres for one of the larger types. This feature of significant travelling may not have been fully appreciated when wild areas were designated as wild cat sanctuaries. Long walks, however, are not commonly considered as a feline need.

Dispersal

Dispersal is the necessary prelude to the solitary existence that characterizes the feline lifestyle. Either in wild species or in the domesticated cat, the young are very social animals until the instinctive process of dispersal occurs. With this process the matured young begin a social detachment from the mother and siblings. They wander away from the maternal territory in search of some exclusive territory of their own, having entered an

Fig. 10.1. Bonded pair looking out of window.

Fig. 10.2. Lions travelling.

asocial phase of life. This complete separation from the maternal base develops, in most species, in the second year of life and reduced sociability occurs at this time.

In the wild species, the dispersing young undertake a search for a new territory that is far removed from the natal region. In the domestic cat, this is not possible with homely constraint and the tie of its food source, but some outreach effort may be made by some, with resulting loss from home in a number of cases. In the wild species the extent of dispersal can be very great, often reaching up to more than 100 km. In its new territory the wild cat is committed to a solitary life except in the lion, in which prides are formed and some male coalitions commonly develop.

The development of dispersal is not closely related to sexual maturation. For example, jaguars disperse at about 18 months and reach sexual maturity at about 2 to 4 years of age. Leopard young disperse at various ages between 14 months and 3 years, but do not attain sexual maturity until older than 3 years. Tigers at the commencement of dispersal stay in the mother's territory for a few months when they are about 17 to 24 months of age, then finally disperse when they reach about 28 months. They do not attain their sexual maturity until they are about 3 to 4 years of age.

Dispersal distances are usually longer for males than females. For example, tiger males travel about 33 km and the females move about 9 km in the course of their dispersal. Pumas spend several months in dispersal, with males travelling about 100 km and females much less. Ocelots spend much time in dispersal, averaging up to 8 months doing so, but they only move about 9 km away from their original home base. Bobcats have an extended and gradual process of dispersal and travel variable distances, sometimes over 100 km. By contrast, in the Canada lynx, both males and females travel similar distances in dispersal, averaging about 160 km. The cheetahs also have similar dispersal distance for males and females, averaging about 25 to 40 km for both.

The dispersal process is very hazardous, particularly for males who usually travel long distances over a long period of time in search of a territory. Females often settle in a location close to the mother's territory. In their wider searches, males may inadvertently invade the established territories of older males and become involved in a fierce confrontation with the territory's owner, which may terminate in injury or death to the more youthful and inexperienced male. In general, the mortality rate among male dispersers is so great that the majority die before finding territories for themselves.

In the case of the domestic male kitten, this inherent instinct of dispersal at the height of development should be anticipated to prevent its loss. It may not occur in every member of the domestic species (after several millennia in domestication) but the dispersal instinct is evidently hard-wired in the wild feline brain and it can be assumed that, as yet another vestigial instinct, it has some residual presence in the domestic subject. No scientific proof exists to show evidence of a dispersal urge in the domestic cat, but an asocial phase in the young animal could be related to it. Sunquist and Sunquist (2002) have provided the above data on this behavioural phenomenon in wild cats, while acknowledging that it is a difficult phenomenon to study. A special search for it in the young housecat might yet reveal it in this species.

Almost like an enforced dispersal, the experimental translocation of wild felines creates circumstances in which the individuals wander about for long periods before settling into a habitat, while some never settle. Instances have been reported where two translocated male pumas, which were identified, returned to their territories of origin. One puma travelled 465 km in 469 days and the other travelled 490 km in 166 days (Sunquist and Sunquist, 2002). These events point to the homing ability of wild felines, in addition to their time spent travelling.

Social independence

The dispersal phenomenon has a role in shaping the feline attitude by preparing the cat for a very independent lifestyle. A solitary existence, as discussed above, requires a related attitude in which the animal preserves its social independence. In the wild, such independence is important to the individual in maintaining its nutrition. Predation is a means of feeding which has a degree of uncertainty – every feed requires a search and great physical effort. The attitude of social independence in cats is an adult development, inherently programmed in the essential feline genome for purposes of survival. Their territory houses food resources and is protected as a private place by the individual feline that acquired it. Social associations would signify a sharing of restricted resources. Among wild cats, starvation is one of the principal causes of death and incidental episodes of hunger assuredly motivate them to preserve sole possession of their sustaining territory. The solitary existence that they lead is the means to that end.

The domestic feline species is undoubtedly possessed of the same socially independent characteristic, although to a lesser and varying degree throughout their population. This explains why most household cats often display an independent social behaviour, in sharp contrast to the typically close relationships that other companion animals have with their human owners. Some household cats do have diminished social independence and typically maintain a more intimate social attachment with their caregivers – unlike the majority in the broad population of common domestic cats.

Territorialism

The possession of a fixed home base is the dominant motivating factor in the life of a cat. In almost every case the ownership needs to be exclusive for each individual feline, for its emotional well-being. Even among social grouping cats such as the lion and the domestic cat, territorial defensiveness is of high importance both within and by the group. Sharing of territory among housecats is initially achieved through tolerance, and by social actuation as the relationship evolves. A cohesive domestic group will bear a strong aversion to any alien adult impinging on the territory. Such an event usually results in fresh territorial marking, especially by any male member in the group. Following territorial damage to the feline psyche, the disruption in customary toileting is likely to persist for a while.

Among the wild species, individual territorial possession ranks as the most important factor in spatial behaviour. While territorial overlapping commonly occurs between one established male and a few separated females, territorial sharing does not normally extend beyond this circumstance. As discussed above, a wild cat's territory serves as its source of food. Loss of this domain may lead to hunger, starvation and even death, if an alternative territory with adequate prey is not found quickly. Such a threat is always present in the feline subconscious, leading to a strong motivation to protect and retain its acquired estate.

In the wild environment, young dispersed males and dispossessed elder males are subjected to territorial impoverishment. A high mortality rate from starvation occurs in these two feline categories. The suffering that attends this form of death is surely severe. This animal has great biological justification in its extreme motivation concerning the importance of territory, either in the wild or in domestication.

Random territorial travelling

Strongly featured in feline behaviour are sessions of random territorial travelling. These are performed on a regular basis, particularly at night. In wild species, the young male undertakes a search for a new territory that is far removed from the natal region. The extent of dispersal in a wild feline can be vast, often reaching up to 100 km. Within its new territory the wild cat is committed to a solitary life, except in such species as the lion, where male coalitions commonly develop. With the indoor domestic cat, homely constraint and the tie of its food source render the search for new territory impossible; some effort to escape may be made, however, often resulting in loss from home. This is a major welfare problem in the ownership of outdoor cats, particularly in urban districts. Full identification of the cat is therefore necessary, either with a tag or an implanted chip.

Miscellaneous Features

All members of the feline family share a large number of behavioural features across species lines. These are core characteristics, many of which exist also in the domestic cat. Apart from the obvious carnivorous aspect there are many others, as follows.

Fecundity

A great contrast lies between wild cats and the domestic cat regarding the probable incidences of young produced in the lifetime of a female. Many wild felines give birth just once per year, amounting from seven to ten litters in their lifetime, with each litter averaging two kittens (or cubs). The domestic cat is theoretically capable of giving birth to about five kittens on 30 occasions during its lifespan, producing 150 neonates. These numbers indicate overproduction in the case of the domestic cat and, in many wild cat species, only enough to maintain its population level.

Littering

Among wild feline species, small litters consisting of one or two kittens are common. The domestic cat often has larger litters, particularly when they are sexually mature. Parturient cats tend to construct a den in a secluded spot for their kittens or cubs. This provides the young with some insulation and protection. Further protective behaviour is shown in the feline habit of actively relocating the young every few days until they begin to walk on their own.

Growth rate and longevity

With small litter sizes occurring among wild cat species, birth difficulties are seldom a problem. A further advantage is the absence of competition between neonates for access to the mammary gland. These conditions favour their rate of growth, thus increasing their survival prospects. The general lifespan among members of the feline family is seen at its best in captive circumstances, where it appears to extend into the teens. In the wild, it is often shorter as a result of sundry hazards. Housecats kept indoors generally have a longer life expectancy than those allowed to roam outside. Seventeen years is a common limit for indoor cats. Free-roaming city cats have an average lifespan of 2 years.

Hiding

All species of wild cats are partial to hiding – from larger cats, competitors or their own prey. It is their *modus operandi*, which is suitable for many of their circumstances. Domestic cats will also frequently hide in suitable places. Without the opportunity for hiding, feline welfare is compromised.

Tree climbing

Tree climbing is a main feature in the behaviour of many feline species. It is an activity they perform proficiently. Perhaps the majority of cats are principally tree dwellers, utilizing tree limbs as walkways and frequent resting sites. Many domestic cats have an urge to climb trees and some will spend resting time in trees (Fig. 10.3a and b).

Swimming

All cats are capable of swimming. Wild species, as those described, do so quite readily; while others, such as the domestic cat, are reluctant unless a life-saving attempt is necessary. Even young kittens of a few weeks old are able to swim. Cats, when swimming, use the dog-paddle motion. Like running, swimming requires a lot of energy and effort, and the cat cannot sustain this exertion for very long. Exhaustion from swimming will ultimately lead to drowning if the animal is unable to reach ground. The antiquated method of eliminating unwanted kittens by drowning was an inhumane, cruel practice. Such an act would warrant the perpetrator a charge of animal cruelty.

Sunbathing

Although most wild cats remain in seclusion during daylight, a number of them are known to emerge from their hideouts around the middle of the day to bask in sunshine (Sunquist and Sunquist, 2002). The manul (or Pallas's cat), for example, has often been observed stretched out on a boulder, or basking on a patch of sand during noontime sun. The sand cat and jungle cat also exhibit this habit. Siberian tigers are known to participate in occasional sessions of deliberate resting while exposed to direct solar radiation, while lions often lie out in the sun on a smooth level surface. The common, house-bound cat is no exception to this feline trait. The domiciled cat is often attracted to rays of sunshine coming through a window in its home, creating a patch of light and warmth where it may rest. Evidently, the cat appears to derive comfort from resting in a sunlit open space – a practice that may aid its well-being (Fig. 10.4).

Tail action

The tail appears to function as a useful appendage, altering the centre of gravity during critical movements to establish stability. It can also signal annoyance by thrashing. Through gentle movement of the tip, equanimity is expressed. It is raised vertically in confident walking and horizontal when stalking.

Sprinting

Quick forms of feline motion can be regarded as prerogative abilities in hunting. The initial dart followed by a short sprint provides instant speed. This brief run is an anaerobic activity, omitting breathing from the action and thereby determining the limit of the sprinting dart and the cat's inability to perform it again without a rest. Door darting can be a problem with some housecats.

(a)

(b)

Fig. 10.3. (a) Tree-climbing housecat. (b) Cat on lead climbing tree; note grasp of trunk.

Meal time

The eating pattern among wild cats contains examples that are applicable to domestic cats. As discussed in Chapters 7, 8 and 9, many are crepuscular in their hunting timetable, indicating two peak periods of eating (morning and evening, approximately). The proper schedule for a housecat's feeding is twice a day – in accord with felines in the natural situation. However, sometimes cats have appetites that do not relate to the twice daily frequently mentioned (Scottish Government, 2010); others manage their intake by having constant access to dry food, with moist food rationed and offered once daily. Some cats are fed best when they call for it, as some can learn to do. In the latter arrangement it is

Fig. 10.4. Resting lions.

assumed that the caregiver is on hand at most times. For every feeding scheme that evolves to suit the cat there needs to be monitoring of its effect on weight and size, since overweight cats are at risk of diabetes. When there is a weight problem, the schedule can be maintained with lesser amounts of feed.

Food variety

An additional feature of wild cat feeding is the wide variety of prey they consume. In most cases, these cats could satisfy their appetites by feeding on their principal prey alone. They do not, however, limit themselves to any singular source of food. For the domestic cat, variety is the best assurance of a well-balanced diet. Moist food should be offered periodically to cats on dry food, for variety.

Food removal

Wild cats often carry their catch away to a different location prior to eating – a habit exhibited occasionally by some housecats if the food is suitable for carrying. Here is another example of an ancient wild gene persisting in the domestic species. Uneaten moist food should not be left beyond the normal feeding time.

Affiliative ability

A large number of wild cat species can show a tolerant, even friendly attitude towards humans. This inherent characteristic deserves more general awareness and appreciation. It is a redeeming feature in a variety of animals often considered hostile at all times. Lions had a tolerant relationship with the African bushmen (Thomas, 1994).

Population welfare

The collective welfare of wild cats is deteriorating seriously, compared with the increasing popularity of the domestic cat. Three human activities have affected wild cats in stages, namely big game hunting for sport, poaching for the collection of fur as exotic garments and the loss of habitat to the agricultural and logging industries. Wild cat populations have been reduced by these activities to very tenuous levels, leading many full species close to extinction.

The current destruction of well-being in individual wild cats is chiefly the result of territorial loss. As a result, territorial overlapping occurs in males, leading to aggressive conflict between the affected cats. Both injured and displaced felines cannot hunt effectively, and subsequently die of hunger.

Genetic inheritance

Originating from this spectacular family, the domestic cat can be viewed as a phenomenon. It has evidently drifted from the wilds of prehistoric times into the homes of modern civilization with great success, continually carrying its ancient genes. Despite domestication over thousands of years, little has been lost from its genetic feline heritage, as evidenced frequently in the comparisons of behaviour in the wild species with the domestic cat.

Selection

As a result of latter-day human selection, additional genetic features are evident in various new pedigreed breeds of cats. Some of these new features are physical, while others are behavioural, and still others combine features of both, as is observed in oriental cat breeds. It is likely that, in some of these cases, physical and behavioural novelties are genetically linked together and originally appeared as mutations from the norm. All pure breeds have special characteristics that have been selected by cat fanciers for their particular attractions. These characteristics dominate older genetic features, which may or may not have been bred out of the breed. It is true that some breeds have lost particular feline features. The Persian cat, for example, has lost a part of its natural excretory instinct, with many individuals failing to make their excretions meticulously in their immediate home territory as performed by other cats.

Mutations

Feline mutations have created major changes such as colour, physique and temperament. There are now many pedigreed breeds that show particularly affectionate behaviour and attraction to human company. Such breeds include the following selection in alphabetical order: Abyssinian, Burmese, Cornish and Devon Rex, Himalayan, Ocicat, Ragdoll and Siamese. Other physically impressive breeds include the following selection: Maine Coon, Norwegian Forest Cat, Russian Blue, Selkirk Rex and Turkish Angora. These breeds illustrate small samples of variation from the common forms of domestic cats.

In the course of selecting original specimens for the new, pedigreed breeds of cats, it is clear that sociable individuals were deliberately chosen and unsociable animals were avoided. In most of these new breeds, the genetic factor of feline social independence has been eliminated. The majority of newer breeds have lost this characteristic entirely, creating increased sociability and popularity with their human associates. The behaviour of sociability in these breeds is to such an extent that their insistent demands for a social connection with their caregiver can become a problem.

General Observations on Comparative Behaviour in Cats

Scratching

Scratching is a key function practised by all members of the feline family. It is most often performed with the forepaws for the following purposes:

- Marking territory as a claim to it. In the wild, longitudinal scars are made by scratching tree trunks with the forefeet and, occasionally, clawing the ground with the hindfeet. Such scratches are also marked on to rock faces. Mark-making is done on a regular basis to serve as a fresh sign of current tenancy.
- Catching prey by anchoring the forefeet into the catch, while maintaining grip during the struggle.
- Manipulating prey as freshly caught or as a carcass.
- Exercising the joints and muscles of the forepaws and forelimbs, by pulling/plucking on the anchored claws rhythmically and alternately. During this exercise the forelimbs are extended.
- Maintaining the claws' points in a condition of sharpness and strength during growth, using the method described above.
- During fighting, clawing (coupled with biting) is performed by either the forefeet or hindfeet.
- Grooming by scratching with the hindfeet.

The claws are normally sheathed, preserving them as sharp hooks. While these sheaths protect the claws from locomotive wear, the cat's paw pads take the animal's weight during movement.

Hunting

Hunting behaviour requires fixed staring and lengthy, concentrated gazing. Even the domestic cat exhibits spells of staring and gazing into space. All cats exhibit daytime resting during wakefulness when they are in seclusion. This is the normal feline method of conservation and self-protection, which is in marked contrast to their activity when hunting. Hunting techniques are similar in all species and involve stalking through cover, slow advancement and then a final leap on to prey. The feline leap may be horizontal or vertical, capable of achieving remarkable distances. Likewise, the domestic cat can leap a horizontal distance of more than 2 m.

The feline hunting procedure can be summarized in a series of connected actions, as follows:

- walking, stopping, observing;
- sighting, followed by a fixed stare;
- creeping and stalking, with a slow forward movement low on the ground;
- freezing and maintaining a perfectly still, low position;
- body wiggle and physical preparation for sudden action;
- darting and rushing forward quickly;
- pouncing, through an upward and forward leap;
- grasping, catching prey with the forefeet and mouth;
- fixed grip by holding down prey while positioned on top of it; and
- killing by biting the neck, either choking or shaking prey until still.

The dart is a remarkable piece of feline kinetics. It begins from a stationary position with a strong forward leap that instantly becomes a sprint on landing. Through this acceleration, the sprint can be maintained for a variable distance depending on the individual animal. The cheetah, for example, can sprint for several hundred metres, while most other species can sprint for only roughly 50 m. The distance of a sprint in the domestic cat and smaller wild cat species is somewhat less, closer to 15 or 20 m. Following a dart, the cat requires a short break to restore its available energy. The glycogen level in musculature demands replenishment, acquired through additional respiration and rest.

Once a dart has commenced, the cat is unable to halt suddenly or change direction. With a darting limit, prey can often escape if it has the ability to shake off the attacker. Many ungulates, such as deer and antelope, are capable of breaking away due to maturity and fitness. As a result, catching such animals may be difficult. Among wild cats, a capture success rate of 50% would be considered average. An equivalent rate of escape suggests that prey are either fortunate or highly fit. In a successful catching attempt the target may be weaker, or natural selection may play a role in the feline's predation.

Overkill

A cat effectively exhausts itself in its chase for large prey, and for this reason it is presumed that a feline will initiate only one kill at a time. This is not to attribute an ethical factor in its killing behaviour. In reality, wild cats will occasionally perform episodes of over-killing. Two examples of such episodes are a lynx's slaughter of numerous domesticated reindeer (Bjärvall, 1992) and a puma's killing of a large quantity of sheep (Sunquist and Sunquist, 2002). In such instances of over-killing, the large predator encounters a number of prey animals in a domesticated restraint, from which escape is impossible. The predator's principal motivation in these cases is not hunger, but some inherent lethal urge.

Vocalization and communication

Cats use a variety of sounds to express themselves or to signal to others. These vary from the Siberian tiger's deafening roar (Vaillant, 2010) to the housecat's silent miaow. Due to intimate contact with their caregivers, domestic cats have much more opportunity than wild cats to communicate vocally with humans. Among the varieties of domestic cats, the oriental breeds are much more vocal than the common occidental types.

In their special, social relationships domestic cats utilize their vocal capabilities to express their emotions, feelings and needs to their human associates. The miaow expression, for example, is conveyed to humans more often than between individual cats. In the latter situation, cats tend to use quiet, short vocalizations for social messaging. During feline dispute, body language is the prevailing form of communication. When vocalization is used, loud expressions of feeling are emitted with harsh assertion, stemming from the back of the mouth. The purr and the greeting trill are expressed with the mouth closed, while various forms of the miaow are made with the mouth partially open (Bradshaw, 1992).

Although the miaow's varieties of pronunciation are commonly deemed to have precisely different meanings, all inflections of it are considered here as attention calls. Some calls of high intensity have preliminary, reproductive relevance, and both sexes emit sharp cries at points in mating. Hissing is an expression of fear or anger and a loud, hoarse cry expresses rage. In a continuing state of pain, there may be suppressed groaning that is similar to purring. The loud roars of large wild cats are expressions of self-identity, intended action or long-range signals of physiological status. Since wild cats live alone except at breeding, complex social vocalizations do not seem to occur. Their vocals during breeding, however, resemble those of the domestic species.

The notable exception to this general state of minimal vocalization in wild cats is found in lions, and among lionesses in particular. With their characteristic social system of living in prides, they have need, among themselves, to communicate at close quarters in their sorority. Females use snuffling sounds to acknowledge others and various other sounds in social situations, variously resembling grunting, moaning and a low chuffing. Lions and other large cats (tigers, leopards and jaguars) are unable to purr – their principal vocalization is through roaring.

Of all wild felines, only the big cats are capable of emitting a hard roaring sound. Their ability to do this is by virtue of a bony structure in the glottis. Smaller cat breeds, however, lack this structure and are therefore incapable of roaring. Some small cats are able to make unusual sounds that are peculiar to their own species. Fishing cats, for example, give out a strange chuckling call, and their kittens are known to make chirruping sounds (short chirping calls are also made occasionally by domestic cats). Furthermore, adult fishing cats, as well as the rusty-spotted cat, the flat-headed cat and the leopard cat are all capable of emitting gurgling noises. All cats, without exception, employ the following five vocalizations: spitting, hissing, growling, snarling and mewing. The first four of these expressions always communicate imperative messages.

Since wild cats (with the exception of the lion) live solitary lives, their means of communicating with others is solely through scent-marking and sign-posting around their territory. Scent-depositing is done through urine spraying, faecal depositing and head rubbing. The function of territorial signs is to indicate the home base of an individual with the given odour. Intruders are left to determine their own response to the territorial sign. A female, for example, will not be deterred by a male's scent, while an adult male will recognize that the territory has been claimed by another male and that deep intrusion will result in an aggressive repulse. When a female is in oestrus, her scent marks will advertise that fact to wandering males.

Mating

The mating procedures of both wild and domestic cats have the same rituals, as follows:

- caterwauling by the male and female;
- tentative male contact;
- oestrous rolling and calling by the female without early acceptance of mating;
- male crouching on top of the accepting female, his legs remaining astride the female's body;
- lordosis by the female with upraising of the hindquarters and turning the tail to one side;
- arching of the back and depression of the male's hindquarters to effect genital apposition and intromission;
- male biting the nape of the female's neck;
- pelvic thrusting by the male for a period ranging from several seconds to a minute;
- outcry from the female as the male begins withdrawal;
- hasty dismount by the male as the female attacks and attempts to bite his head;
- a resting phase averaging 5–15 min by both the male and female;
- repetitions of the above rituals by the same male; and
- the further involvement of another male, or several males, in the mating process as oestrus continues for a few more days in the accepting female.

Oral actions

Sampling items that may be edible and drinking are performed regularly by all species of cats. Prior to grasping food, the cat will usually inspect it through sniffing and tasting. During consumption, flesh is licked from bones with the aid of the tongue's rasping papillae.

All felines have very purposeful canine teeth, chiefly utilized for spearing prey. During chewing, a cat tilts its head slightly to one side, using the carnasial tooth on this side to slice food for swallowing. The canines are well placed for this purpose, located at the front of the mouth as opposed to behind a protruding arc of incisors, as is typical in other mammals. The short face, so characteristic of cats, allows the jaw to have a firmer bite. The cat's smallish incisors are situated in a straight row between the canines. The canines are essential catching equipment; to such an extent that damage or loss of one drastically reduces the cat's ability to hunt, leading to starvation and death.

Yawning

Yawning often occurs as an integral component of the compound stretching exercise of pandiculation (*Dorland's Medical Dictionary*, 2003). This oral action in cats is usually performed during periods of inactivity without the stretching of other bodily parts. As a singular exercise, feline yawning is performed with particular vigour, causing the mouth to be opened so that the upper and lower jaws form a 90° angle of separation. This extent of mouth opening is not observed in most other animals, but is a characteristic of felines from the largest wild cats to the common housecat (Fig. 10.5).

One function of yawning is to fully articulate the temporo-mandibular joints, stretching their related muscles. This exercise has the effect of maintaining optimal efficiency in mandibular joint function during periods of oral inactivity. The daily incidence of yawning in cats, both wild and domesticated, is higher than commonly occurs in other animals. Evidently, the capability of extreme jaw extension in felines is important to maintain. Yawning also reveals the spacious mouth of a cat, with its highly arched hard palate.

The relationship between wide yawning and vital life is apparent in the feline manner of predation. Most wild cats include in their prey animals that exceed themselves in size. After catching such prey, the feline method of administering the *coup de grâce*, or executing stroke, is by gripping the prey at the junction of its head and neck in a wide bite, throttling the animal and causing a prompt death. The encompassing bite requires an exceptionally wide expansion of the jaws. This is rooted in the feline genotype as a special piece of behaviour. The related yawning in a cat serves as an indication that the animal is in functional health or in a condition of well-being.

Social licking

Licking is a common activity in all species of cats. It is frequently used in maternal care and as a manner of ingestion. Maternal licking is performed on new kittens for cleaning and stimulating excretion. As the kittens grow, this activity is continued – maintaining the maternal bond through tactile caressing. Mutual licking periodically involves exchanges between a breeding pair, particularly in the courting phase where it apparently facilitates mating.

In its social use, licking takes places quite frequently between individual cats in the species which form groupings. Members of a lion pride, for example, engage in licking one other without reproductive connotation or gender basis. Female lions engage in this activity with each other, and if a male is permanently included in the structure of the pride it may lick the surrounding females. In an assembly of feral domestic cats, licking is common between associating individuals. Sometimes, one individual will perform more social licking than

Fig. 10.5. Front view of domestic cat yawning.

others in the feral group. It evidently serves to maintain harmonious dual relationships that, in turn, collectively perpetuate a peaceful state among the group's members. Social licking in a feline group has an important function for the group's cohesion (Fig. 10.6).

Even in a household setting, allogrooming between partnered cats takes place in licking sessions. Commonly, one individual cat will begin by restraining the other cat with one paw and energetically licking some portion of the associate – often about the head. At times, two paired cats may attend to each other simultaneously in joint grooming sessions. Such allogrooming is regarded as social grooming, although it also serves as a cleaning practice. Sessions of allogrooming may be of varied duration, and one grooming partner may cease cleaning while the other continues.

Lookout posture

The cat's daily levels of mental consciousness cover a spectrum of states from nil to maximal. These include long periods of rest, light sleep, catnaps, light somnolence, drowsiness, general awareness, specifically focused attentiveness and a lookout state of watchfulness. The latter is contained in a posture of sitting upright on the chest or haunches with the head elevated and the tail outspread (Fig. 10.7). The head may be swivelled to nearly 180° in response to any auditory stimulus out of its forward view, without movement of the body.

This watchful state is adopted at random and is maintained for various periods of time. During this state, the disposition of the cat is one of neutral motivation but with full awareness relating to the immediate environment and events within it. It is a

Fig. 10.6. Social licking.

Fig. 10.7. Lookout postures.

useful time for making contact with an otherwise withdrawn cat for any given purpose, such as medication or grooming.

Affection

Within the wild feline population there is abundant evidence of affection in families. A good example is observed in a lion pride. In other species, there is affection between a mother and her litter, and vice versa. Litter mates also express closeness to one another. It is only when the young males achieve sexual maturity that they develop other motivations and depart from the family scene. Young females leave in search of their own territories as well, but these new regions are often not far removed from their natal origins (Sunquist and Sunquist, 2002).

As previously mentioned, many wild cats have a friendly disposition towards humans, in spite of a common misconception to the contrary. Several species can be domesticated without any period of adjustment to human possession or domesticity. Prime examples include the ocelot, the caracal and the serval. As a further example, the Bengal breed of domestic cat has evolved from breeding natural domestic cats with the wild Asian leopard cat.

Among the larger cats, the cheetah has a long history of being kept as a pet. Only recently has this practice of domesticating cheetahs been made illegal. Within the feline family there is a strong element of affiliation with caring humans. This characteristic deserves better recognition, in order that such cats can be regarded as friendly, without the intent to harm. Exploiting the friendly nature of a wild cat through domestication, however, may not ensure its true well-being. The domestic cat has a great capacity for affection even with an alien species (Figs 10.8 and 10.9).

Human attack

In marked contrast are species that have a record of human attack. The lion, tiger, leopard, puma and jaguar have been known to show aggressive pursuit of humans, sometimes as prey. The puma and jaguar have a history of occasional attacks on people intruding into their territories. The most deadly felines are found, however, among lions, tigers and leopards. The Bengal tiger has the worst reputation for such savagery, while the remote Siberian tiger is also guilty of such acts. A factual account of a remarkable case of such 'man-eating' in a Siberian tiger (Amur tiger) has been detailed in a recent book titled *The Tiger*, by John Vaillant (2010), who describes the natural history of this huge, ancient cat that is under the threat of extinction from chronic and continuing human conflict.

In addition to its present-day African habitat, the lion once had a well-established Asian existence, particularly in India, but became ousted from most of that region. The Indian government began a programme of re-establishing the Asiatic lion in the region of the Gir Forest, in the State of Gujarat. When the lions were brought in, the majority of the local human population were relocated away from the area. The lion group became established in the new territory and started breeding, and as their numbers increased, so did territorial overcrowding. Excessive fighting resulted. Consequentially, many lion males became displaced from their

Fig. 10.8. Feline affiliation with canines.

Fig. 10.9. Tentative greeting.

habitats and began to wander beyond the area set aside as their sanctuary. Individual males began attacking and even eating humans they encountered, despite the availability of their natural ungulate prey. These attackers were then transferred to a different location, but still remained dangerous. The only solution to ensure the safety of the people was to destroy any lions responsible for inflicting such harm.

The cause of this practice of man-eating is unknown, but it is a cataclysm in predatory behaviour that appears to be associated with a reduction in the cat's well-being from injury or loss of territory. It can be concluded that a loss of well-being results in this extreme form of delinquent predatory behaviour in big cats.

Maternal functions

Kittens (or cubs, in the case of big cats) cannot live without maternal care. This is an obvious fact in the immediate or short term, but it is also a fact in the essential development and instruction of the maturing juvenile cat. In her extended maternal role, the mother of a kitten must begin by suckling and protecting her young.

The majority of wild cat mothers provide protection to their litter by transferring their young from place to place at intervals. Infant wild cats are vulnerable to a range of potential predators, including adult alien males of their own species. By periodic removals, the litter is less likely to be tracked down by investigating predators using established, scented clues. Additionally, the mother will fight off any intruder aggressively, as has been described among lions. If one member of a litter is lost to a stronger attacker, the mother will usually escape immediately with any surviving kittens.

Weaning

Most wild cat mothers suckle their kittens for lengthy periods such as 3 months. The stress of weaning is eliminated by bringing the young solid kill portions of prey even before they are able to eat it. This way, they become familiar with the material for which an appetite will soon develop. The mother maintains this food supply until the kittens are able to travel with her during hunting trips. At this time, the young animals learn from their mother's performances the techniques of hunting and killing. The kittens are undoubtedly born with these instincts, but the practical techniques of predation have to be taught to them by their mothers in their post-weaning stage. Wild cats raised in captivity do not receive this education.

Maternal connection

When the young have learned to hunt and sustain themselves, the mother still has maternal duties. She will allow them to share her territory, and she will not come into oestrus while they remain in her charge. Young males at the approach of sexual maturation, or before, will leave the maternal scene of their own volition, while young females will not leave their natal region. They may remain in the company of the mother for extended periods or even permanently, as is the case with lion societies. By remaining adjacent to the mother's territory, the female offspring may wander into the extensive territory of their father. No cat, however, has a certain paternity, given the style of free breeding that is the feline norm.

11 Abnormal Behaviour and Training

Stress

The term 'stress' originated as a commonly used word to indicate some undue pressure on an object or person, resulting in some form of reduced soundness. It still has that meaning in common language but, additionally, the term has been adopted as a physiological factor with more exact meanings. Various modern definitions have been advanced, suiting different circumstances and concepts. Here, the term is used with the following definition: 'a disturbed state in the animal's mental well-being from an environmental insult'. While it is acknowledged that, with broader use, there can be physical stress as well as mental stress, the latter type is more relevant in this text since the focus here is on the cat's mental well-being that is evidenced in behaviour.

The cat relates intensely to its environment, both in the wild and in domestication. Its welfare is rooted in its established environment that provides it with physical security and food. Alterations in its environment can be very disturbing to the cat, again in the wild or in domestication. Adverse environmental alterations can be, for example, weather, loss of prey, intrusion of a competitor, displacement from home, change of ambient features or addition of new associates. These changes in its environment can cause stress to the normal cat that depends on consistency in its ambience for security and safety.

With mental stress, as a form of suffering, the cat's behaviour shows various forms of alteration. Among these are aggression, soiling, chewing, excessive scratching, hiding and anorexia. Clearly, mental equilibrium becomes disturbed in the affected cat. The stressful factor is absorbed into the conscious state of the cat's brain and creates excessive neural activity in the limbic system which is relayed to the hypothalamus onwards via the pituitary and other endocrine glands, notably the adrenal cortex, motivating unusual actions, as mentioned below (Siegel and Edinger, 1983; Klemm, 1984; Siegel and Pott, 1988; Dennett, 1991; Carlstead *et al.*, 1993; Eagly and Chaiken, 1993; Siegel *et al.*, 1999; Nelson, 2000; Gregory, 2004).

Common Behavioural Problems

Bessant (2004a) has reported on a survey of data on features of cats' lives obtained by the Feline Advisory Bureau using questionnaires returned by the owners of 1853 cats in the UK. Of that number of cats, 1389 had shown a behaviour problem at some time in their lives. With roughly three out of every four cats having a behavioural problem during their lifetimes, it is apparent that many cats encounter mental/emotional difficulties in their domesticated existences. Some of these can be eliminated or prevented by neutering (Knol and Egberink-Alink, 1989). The Feline Advisory Bureau survey revealed seven principal behaviour problems as shown in Table 11.1.

In addition to these seven types of behaviour, there were some less common ones noted as disorders, such as eating wool, stroke and bite (Jekyll and Hyde reaction), excessive nervousness, nocturnal vocalization and nervous reaction to noises. Many of these disorders stem from the natural feline genome and, on the whole, they tend to be created by environmental factors.

Significant differences are reported between the genders in the frequencies of some of the above conditions. For example:

- inappropriate indoor toileting, more common in females;
- indoor spraying, more common in males;
- cat-to-cat aggression within the home, more common in females;
- cat-to-person aggression, more common in males;
- fear of certain sounds and sights, more common in females; and
- outdoor fighting, more common in males.

Table 11.1. Common behavioural problems in felines (data from Bessant, 2004a).

Behaviour	Number	Percentage[a]
Inappropriate excretory conduct	709	38
Over-grooming	159	8
Cat-to-cat aggression	202	10
Cat-to-person aggression	103	5
Separation from caregiver problem	107	6
Destructive scratching	75	4
Inappropriate sexual activity	34	2

[a]Out of 1853 cats in total.

Featuring in the incidences of any of the conditions given above is the factor of individuality. Some cats simply adapt more readily than others to the environmental circumstances underlying many behavioural problems. For example, some cats have difficulty adapting to an indoor life, while others accept that lifestyle very readily. In fact, in the Feline Advisory Bureau study it was found that 45 cats that were free to go out never did so. However, one-third of all cats that were once allowed out, but then retained indoors, continued trying to get out. Cats can adapt better to consistencies in their manner of husbandry and the latter, in the case of indoor confinement, should have quality of interest with toys, a partner, resting shelves, scratching post, good toileting facilities, good feeding and good animal care along feline lines.

The fact remains that individuality is of a very high order among cats (Mendl and Harcourt, 1988). It is common knowledge among cat keepers that no two cats are exactly alike in personality/character. The one psychological feature that they do share is a complicated mentality. It is vital to realize that a very sudden change in a cat's personality is often due to the commencement of an illness or pathological condition (Fogle, 2011), just as it is not unusual for any animal's behaviour to alter due to a pathological condition (Broom, 2006).

As the veterinary pioneer on cat behaviour, Michael Fox (1974) was the first to emphasize the importance of early socialization of kittens in order to make them comfortable and manageable as pets in later life. This fact is still emphasized by contemporary experts on cat behaviour such as Hotchner (2007), who specifies that the optimal period for kitten socialization by handling is 2 to 7 weeks of age. Hotchner states that kittens that do not receive social handling become cats that are predisposed to fearfulness. Also predisposed to fearfulness are the 25% of cats that have a fearfulness gene passed on to them by their parents, according to Hotchner.

As regards veterinary attention to abnormal behaviour in cats over the years, leading veterinary authorities have concentrated on the issues of aggression, spraying and soiling in housecats (Voith, 1979; Houpt and Beaver, 1981; Hart and Cooper, 1984; Marder, 1991; Cooper and Hart, 1992; Manteca, 1995; Houpt *et al.*, 1996; Houpt, 1997; Schwartz, 1999; Hunthausen, 2000; Mills and White, 2000; Mills and Mills, 2001; Pryor *et al.*, 2001; Beaver, 2004).

Separation anxiety

The term 'separation anxiety syndrome' (SAS) is given to housecats that engage in disturbed behaviour after being left alone in the house by the caregiver (Schwartz, 2002, 2003; Hotchner, 2007). When the residency of the accommodation consists of only one person plus the cat, this condition is more likely to occur. (A comparable emotional disturbance is recognized in dogs.) The feline state occurs when a close bond between the caregiver and the cat appears to be broken. The responsible factor is the apparent rupture of the human–animal bond that has become widely recognized as an emotional link between a person and a pet. Separation anxiety is most common in cats that are either very young or very old. The very young and the very old are more emotionally vulnerable than cats in other age groups.

The condition resembles a sudden estrangement that causes distress; in fact, the term 'distress' is more suitable than 'anxiety' for this condition. Anxiety is a mental state that may not have behavioural signs, while distress implies a disturbed mental state that has very evident behavioural symptoms. Certainly, the condition has disturbing signs in affected cats. They engage in destructive behaviour when left alone. This can take such forms as chewing wires, damaging furniture, knocking down objects, accessing cabinets and disturbing the contents, and excessive mewing. The distressed behaviour in some cases progresses to self-mutilation. The term 'separation anxiety syndrome' is well established internationally and needs no change of label for the condition, but the degree of emotional disturbance to the affected cat should not be minimized, since it is a form of significant suffering. Anxiolytic medication may be needed in extreme cases. Such treatment would not provide a cure, merely temporary relief.

The course of treatment advised by Lachman and Mickadeit (2002) features such tactics as making departures discretely, pretending to leave at times but not doing so (so that the cat becomes desensitized to the usual departure clues), providing regular sessions of friendly interaction at times unconnected to any departure, ignoring the cat for occasional periods while at home and refusing lavish attention to the cat on returning home. These expert cat trainers strongly advise against any punishment to the cat for destructiveness, since this is counterproductive.

Apart from the queen with a litter, close social connections are not prominent features between cats in feline sociobiology. Innumerable exceptions do occur, however, and feline bonding with a person, or another animal, can easily exist under favourable conditions. In such an instance, an assumed breach of the bond can be an emotional experience that is disturbing to the cat when it is suddenly deprived of the caregiver's company and left in isolation. The altered behaviour with this condition can occur in such activities as wire chewing, excessive scratching and tearing furniture, etc., as mentioned above. This may be anticipated and avoided by placing the cat, during a temporary absence, in a secure room with an abundance of its play items and a softly playing radio or even television (Fig. 11.1).

Another tactic that can be quickly effective, if carried out carefully, is simply to get an additional cat as a companion for the one with the separation problem. Such an addition should be a much younger female, neutered and introduced in a cautious manner that does not immediately antagonize the principal cat.

In a number of separation anxiety cases the author has personal knowledge of, the problem stemmed from the manner of the cats' care when they were kittens. In each instance, the kitten received a very high level of ESH during the second month of age and into the third. This appears to correlate with an acute separation anxiety that was subsequently shown in later ownership of these kittens when they became cats. Perhaps there is a hidden danger in very effective ESH, by making such kittens predisposed to strong dependence on constant caregiver presence.

Fearfulness

Domestic cats are predisposed to fear. Mostly they are ready to go into a state of fear in response to surprise, being caught or restrained, sudden change of location, experience of violence or perceived threat of it. Their fear is exhibited in various displays or forms of behaviour. The classical display of fear in cats is the state of fight or flight in which the animal stands in an oblique position to the source of threat, with the back highly arched, the feet close together, hair raised in piloerection, eyes staring, pupils dilated, the mouth partly open and

Fig. 11.1. Lone cat watching another cat on television.

emitting a cry. This posture is maintained briefly, usually becoming resolved by slow withdrawal while watching the source of alarm, or running swiftly away from it. Occasionally, this state of fear changes to rage (as discussed below) and the cat then threatens or attacks the stimulus with an aggressive act of biting and scratching.

A more common form of behaviour in fear in cats is hiding in a concealed location and adopting an immobile, crouching position. Such a reaction occurs if the state of fear is persistent, for example when the animal is experiencing loss of its home/habitat, or is in a threatening environment. Hiding places are often under doorsteps, below bushes, in sheds, under cars, in empty buildings, in cupboards, in basements, in false ceilings, etc.

Entrapment

A state of entrapment can follow concealment, or it can result from investigation/exploration. A state of fear develops in the latter situation also. The animal is then imprisoned and will call out regularly, sometimes with a strange tone that spells out distress. The sound is a drawn-out howl from the back of the throat.

In retrieving a cat from a hiding place, it is usual to find that the animal is reluctant to be moved. The advised procedure is then to open up the concealment and allow the cat to emerge in its own time. In the case of entrapment, however, it is essential to rescue the cat as quickly as possible. The trapped animal may have been held there for some time, even weeks, without food or water. The procedure is to remove the facing of the entrapment after confirming that the animal is immediately behind it. This can be ascertained by making a hole, near floor level, to hear that the cat's calling is exactly there. The cat's fear will end immediately it is freed. It will need sustenance at once.

Aversion reaction

As the result of a traumatic experience, a cat can learn instantly that the essential feature of the experience must be avoided thereafter. The nature of the trauma may be physical, such as a burn or injury, or psychological, such as a fright. The trauma will have greatest effect if it is both, causing shock. The traumatic event will act as a lesson from which the cat will react (either rationally or irrationally) to avoid the main, identifying feature of the aversive event, or the place where it occurred.

The avoidance reaction results in immediate changes in the cat's behaviour, such as a general loss of confidence and uncertainty about its ambient security. Initially, the animal is likely to hide and to be in a state of torpor while remaining in one location for an extended period. In addition to the dressing of any lesion, the best management would be the provision of warm, comfortable and quiet seclusion, with a supply of fresh water. On its emergence from hiding, the cat will show a reduction in its normal movements and wanderings. Recovery is delayed and behaviourally is often incomplete, leaving the animal with a permanent aversion to the principal feature of the experience and a change in its personality.

Aggression

Aggressive behaviour in the cat has a history of fundamental study. The disorder is deep-rooted in the limbic system. Adamec *et al.* (1980a) were the first to show the limbic control of such behaviour. Maeda and Maki (1987) subsequently indicated the limbic's hypothalamic section to be involved in aggressive behaviour in the cat, with Siegel *et al.* (1997) further implicating the limbic's amygdala with additional involvement in such feline aggression. With the basis of such aggression located in the limbic system, it is evidently an emotional condition. Schwartz (1998) reported on the type of situation where this aggression may be triggered by emotional circumstances. Cats often show aggression to each other by using ritual displays of body postures and throaty vocalizations (Morris, 1987), but many of these situations become resolved without a real fight developing. Some cats, especially toms, are prone to fight others viciously. Any such cat should be neutered.

Clinically, cases of aggressive behaviour occurring in cats are not always of similar type. Chapman and Voith (1990) first described 14 cases of cat aggression directed at people, as a different type from other forms of abnormal aggression. Later, Chapman (1991) provided a classification of different types of feline aggression, shedding more light on the general problem. The forms of aggression now recognized are: predatory aggression, play aggression, territorial aggression and temperamentally based aggression (Fogle, 2011). In addition to different types of aggression, differing varieties of circumstances are recognized as being associated with episodes of feline aggression. These include any painful condition that the cat is experiencing, fear in a given situation and, of course, the 'Jekyll and Hyde reaction' as termed by Frazier and Eckroate (2008) and Lachman and Mickadeit (2002). This reaction is also termed 'stroke and bite' or 'petting aggression'. It results from overstimulation of the integument by stroking and is discussed later in this chapter.

In its worst form, aggression in domestic cats is an antisocial disorder and Crowell-Davis *et al.* (1997) made this abundantly clear in their clinical account of the condition. The management of feline aggression is largely dependent on proper differential diagnosis of the condition. Frank and Dehasse (2003) have shown that this is important practice in case management. Cat-to-cat aggression, in which the animals may hiss or growl at each other before any fighting develops, can be arrested by water-spraying the pair, or the aggressor, as soon as an aggressive attitude is seen developing. Animosity may continue between the cats involved, but enactment of aggressive behaviour is likely to be inhibited by such training if carried out assiduously and consistently.

A cat that is terrified of another may, on encountering it, show a typical display of crouching down with teeth bared and ears back while spitting or hissing. Anyone witnessing this display directed to them should not attempt to face the cat down, but immediately stop looking at the cat and withdraw. Reading a meaningful posture, such as this one, is important in cat management. Prompt defusing of a threatening display, by immediately taking eyes off the cat and slowly moving away, is the advised course of action. However, should the cat attack the person, it must be rebuffed immediately and the person attacked must defend him or herself with some form of shield, such as a cushion. The cat must then be subdued, for example by covering the animal with a coat or blanket. The animal can be restrained under a cover until it has become quiet. In any event, the cat cannot be allowed to win the encounter since that would encourage it to repeat the incident (Fogle, 2011).

While some veterinary practitioners have approached the problem in a variety of ways (Voith, 1979), no definitive treatment has emerged, beyond castration of toms (Heidenberger, 1997). Unfortunately, castration of toms after they have attained puberty does not always change patterns of behaviour that were acquired when the cats were entire. Adamec *et al.* (1980a,b) first showed that early experiences in the young cat's life can implant aggressive behaviour. The route to effective management of abnormal feline aggression probably requires a combination of approaches involving neutering, training, therapy and environmental control, such as giving the animal a neutral environment, without any feline competition or stressful ambience.

The feline has the capacity for prompt aggression in its nature (Adamec *et al.*, 1980a,b). Aggression is a force for instant action in capturing prey. The aggressive state is brief under such circumstances, subsiding quickly after the predatory action is completed. Instances of cat-to-cat aggression can be a problem, but these are essentially normal in natural matters of conflict (Manteca, 1995). The Burmese breed of domestic cat and wild cats in territorial disputes show that aggressive incidents are normal components of feline life. Sometimes cases of cat-to-cat aggression become problems to the caregiver in multi-cat households. These cases are usually the result of some anomalous social situation with the cats, such as frustration, crowding, competition or inappropriate mixing. These are common mental stressors for cats (Siegel and Edinger, 1983; Hart and Cooper, 1984; Glusman, 1985; Siegel and Pott, 1988; Siegel and Schubert, 1995; Siegel *et al.*, 1997).

Creating a realistic social situation that is in line with the realities of feline dynamics can often resolve the problem of stressful circumstances in domestic containment. For multi-cat units, measures such as decompressing the feline assembly, expanding their individual territories and eliminating sexual conflicts can be very effective in dealing with social stress. Isolation of an affected cat can be therapeutic since it is not a problem for most cats and is in accord with general feline modes of living. The principal cause of aggression in domestic cats is social stress and isolation can relieve this if a good standard of animal care is maintained.

In some cases, domestic cats can show a syndrome of aggression directed at people (Beaver, 2004; Schwartz, 1998). Companion animals do not normally have aggressive dispositions focused on their keepers, but this can happen occasionally (Marder and Marder, 1985). The regular occurrence of aggression towards a human can be regarded as definitively abnormal. An aggressive disposition in a domestic cat has its basis in the emotion of anger linked to an underlying state of readiness to attack another subject, particularly an alien subject (Glusman, 1985; Hassanain *et al.*, 2003; Knowles *et al.*, 2004). A cat that is predisposed to aggressiveness has an affective disorder primed with a setting of neural and hormonal chemistry that has been established genetically or stressfully, or both (Siegel *et al.*, 1997). With such a system of roots, the abnormal behaviour of antisocial aggression in a domestic cat can be seen as difficult to control or eliminate (Lachman and Mickadeit, 2002). Experimentally, noradrenaline administration is effective in treating affective defence behaviour in cats (Barrett *et al.*, 1987) and the administration of dopamine can control quiet biting in cats (Siegel *et al.*, 1997).

The syndrome of abnormal aggressive behaviour against the caregiver of domestic cats occurs in various forms of expression (or degrees of affliction) and in two types of release. At one extreme form of expression, the cat may attack the person by vigorously biting and scratching the limbs or face; at the other extreme, the cat may only bite mildly at the hand or ankles of the person. The release of transient, aggressive behaviour can follow some innocent provocation by the person (see Jekyll and Hyde reaction) or it can occur spontaneously in unprovoked anger (Reisner *et al.*, 1994). Unprovoked aggression can occur as a very rare, psychotic condition with unknown aetiology which may or may not be controlled by lengthy, specialized training. Such extreme aggression can occur in psychotic episodes that indicate a profound and enduring disorder that can warrant euthanasia.

In the course of controlling an aggressive cat, which may have a dominant disposition (Krotewicz and Romaniuk, 1995), the caregiver should use firm self-defence against the cat if the animal is making an attack on the person. To train a cat with an aggressive tendency, the trainer can use a water pistol or large syringe to direct a strong jet of water against the cat when it attempts an aggressive move. A series of such water-jetting measures is a good method to inhibit a cat from performing any improper act. This treatment is non-aversive and harmless, but it must be done in timing with the unwanted act by the cat, to associate the punishment with the action. The latter is a key to effective training (Voith, 1979; Marder, 1991).

The Jekyll and Hyde reaction

Many housecats show a fierce reaction to the stroking of their back or flank, as it can suddenly become excessive stimulation. Technically, the cat is highly kinesthetic – a factor that is basic to the cause of this reflex. The response has been termed the 'Jekyll and Hyde reaction' (Frazier and Eckroate, 2008), also known as the 'stroke and bite syndrome' (Feline Advisory Bureau, 2009). The Jekyll and Hyde reaction is a suitably expressive term for this sudden change in behaviour; it can be observed in a cat that appears to enjoy its handler's attention, only to suddenly become aggressive and tear into him or her. Once this reaction is induced, it is promptly repeated if further stroking is immediately attempted. Attention to the cat should therefore be ceased, and the animal's usual temperament then returns to a neutral state.

It is not suggested that the Jekyll and Hyde reaction should be sought as an indication of normality – on the contrary, it ought to be avoided. It is, however, a common reaction that needs to be understood and appreciated as a special feature of the cat's phenomenal aggressiveness resulting from stimulation beyond a threshold of sensitization. In addition, the Jekyll and Hyde reaction resembles the female's reaction to the male immediately after mating has occurred. This latter reaction often causes scratches or tears to the ears, and bites to the neck or cheek of tomcats.

In the event that a handler encounters this reflex, the hand should be quickly withdrawn to avoid injury. The reflex does not go beyond the initial reaction and the cat makes no further attempt to attack the handler once the hand is drawn back. If a bite or scratch occurs, the wound should be washed immediately and treated with a suitable antiseptic. The wound may be of a puncture type, in which case an effort should be made to make the treatment deep enough to reach the depth of the wound. The danger of contracting cat scratch fever should be realized.

The Jekyll and Hyde reaction must be understood as a natural, instinctive action and not a calculated assault on the handler. It is as much a part of feline expression as purring, miaowing or licking. Punishment should not be applied to a cat in an instance of this reflex and the handler's attitude to the animal must not be diminished by the experience. If or when it occurs, this reflex should be regarded as a discovery of one of the mysterious idiosyncrasies of feline nature.

The sudden aggression caused by this reaction can occur in other situations where there is precise and immediate bodily contact imposed upon the cat. The reaction evidently arises when a stimulus of physical contact threatens or exceeds the limits of the cat's sensorial tolerance. The cat's skin is well supplied with sensory receptors and this animal may be subject to integumental hypersensitivity. One type of mechanoreceptor is located in groups as touch corpuscles enclosed in Merkel cells (Iggo, 1982). These structures, conjointly with C-mechanoreceptors, are particularly sensitive to skin stimulation, particularly in the dorsal and posterior regions of the cat's body. When there is excessive stimulation of these units, the animal responds with brief aversion.

Repeated stroking of the cat along the lie of its hair can cause a build-up of static electricity in its coat. Such an electrical charge can be eventually conducted from the coat to touch the animal's skin. This apparently causes a brief experience of pain in the cat causing it to react with sudden aggression. Proof of this electrical factor can be observed in some cases by appropriate stroking of the cat in darkness, when a small bright spark can be seen coming from the coat at the end of each stroke. Evidently the build-up of static electricity in the cat's coat from recurring friction is the basis of the Jekyll and Hyde reaction.

Indoor spraying

The other enduring behavioural problem of great annoyance to any cat keeper is indoor spraying, usually by male cats. This also has a long history of study (Blackshaw, 1992; Horwitz, 1997). Usually there is an environmental cause notable in the development of the problem (Dehasse, 1997). By the application of a synthetic facial pheromone (Mills and White, 2000; Mills and Mills, 2001), control over indoor spraying by cats can be effected. Today, with the use of the anti-stress pheromone Feliway, as a spray applied strategically at spots where urination has occurred, the problem can be brought under control (White and Mills, 1997; Fogle, 2011). Eliminating the frustration of any competitive cat is a vital factor in treating this behaviour. Bearing in mind the fact that most behavioural problems in domestic cats have an environmental basis, it is essential in addressing any such problem to take account of environmental changes that require to be put into effect. Extroverted cats are reported to vent distress or frustration on to furnishings, while introverted cats are more likely to adopt compulsive behaviour, such as over-licking, or develop bodily symptoms (Lachman and Mickadeit, 2002).

The excretory problems given above are, in effect, essentially symptoms of mental disturbances in the cats. For example, Wright and Amoss (2004) reported that house soiling and aggression were problems in kittens in their first year after being adopted from shelters. This indicates that such abnormal forms of behaviour are associated with the emotionally demanding circumstances of adjusting to new accommodation. These forms of behaviour, as symptoms, indicate a state of mental stress that can be termed dysphoria. Treatment with an anxiolytic drug, such as cloripramine, has been helpful in some cases (Fogle, 2011). Detomidine was used for tranquilization in cats, following clinical reports of this drug's effectiveness in cats by Young and Jones (1990) and Verstegen *et al.* (1991), but the anti-anxiety drugs in use now are diazepam, lorazepam and alprazolam. Other anti-depressant drugs in use for cats are fluoxetine, sertraline, amitriptyline, clomipramine and imipramine (Lechman and Mickadeit, 2000). No such drugs can remedy the problem alone, although they can make a nervous cat more amenable to management; any such drug can only support diligent training.

Training a cat to use its litter box can be done by placing the cat in a pen in a quiet room with the litter box in the pen. The cat is kept there until it has used the box. It can then be rewarded by some means. The procedure is done immediately after the cat has eaten a large meal. The caregiver remains in the room containing the pen to reassure the cat and to take it out of the pen after the animal has performed satisfactorily. This may require some repetition until the cat has learned properly. The litter box can be kept near that site for a period after training is complete, then removed to a quiet place away from household traffic and its feeding area. The full procedure is described by Fogle (2011).

The dysphoric excretory syndrome

The feline's concern over its territory is a great mental demand on the animal. The behaviour among wild cats shows the imperative nature of territorial ownership. Fights to the death are not uncommon between wandering males and residents over boundary trespasses. The intensity of territorial possession is greater among male cats than females, but all cats relate to territory as a life source. Particularly in a multi-cat house, if a male domestic cat has a territorial challenge that it is unable to resolve, a mental disturbance is likely to occur, especially in uncastrated, or late-castrated, males. The late-castrated cat will have experienced the attitudinal features and motives that are primed by testosterone and some reserve, or memory, of this state can persist after castration.

The mental disturbance from territorial frustration can be of a high order in some individual cats, leading to acute dysphoria, a mental disorder that affects conscious behaviour adversely. With this development, the cat performs anomalous excretory behaviour within its household. This takes forms such as leaving faeces uncovered in unusual sites in the home and spraying urine actively on upright surfaces, a syndrome of behaviour that is suddenly out of character for the particular animal. Disciplinary measures to stop these actions merely frustrate the cat further and are ineffective. The disorder is likely to continue, leading in many cases to a decision to have the cat permanently removed from the household, possibly euthanized. The condition can therefore be fatal for some cats.

The aetiology in this disorder can be seen in two aspects. First, the cat has received a mental insult from some radical change in its territorial property, leading to a mental effect. Second, the affected cat reverts to its genotypic behaviour in which the defensive norm is to defecate on the boundaries of its territory and to spray urine on these sites and on prominent, upright structures in its territory. These are the normal feline tactics to declare exclusive ownership of territory. They are also aggressive signs, serving as a warning to any intruding cat that defiance of the signs will lead to a conclusive fight.

It will be apparent that the syndrome of dysphoric excretion in a domestic cat is a reversion to the natural method of territorial marking used by the wild feline family, particularly entire males. What is needed in the treatment of this condition is a clear understanding of its aetiology. An uncastrated cat would require neutering and action taken to provide accommodation that is an exclusive territorial possession. This fresh accommodation would need to be entirely free of feline sights, sounds and odours. The manner in which such isolation could be established would depend on the available environmental accommodation. Even restricted space, for a spell, would be preferable to disposal.

Frazier and Eckroate (2008) stipulate that any form of excreting outside the litter box by a house-kept cat is a symptom of something gone wrong. This is an important point. Treating the symptom alone is not adequate in dealing with the basic problem that may have multi-factorial causation, as these authors suggest. They list no fewer than fourteen possible contributing factors, posed here as pertinent questions covering the main points:

- Might there be a medical or physical disorder?
- Does the litter box have a proper location and continuous accessibility?
- Is the number of boxes adequate for a multi-cat home?
- Is the litter box of suitable type, size and carefully cleaned?
- Is there a stressful environmental situation affecting the cat?
- Did the cat ever learn to use a litter box previously?

A positive answer to any of the above enquiries could be the clue as to the cause of the cat's wayward excretory behaviour and guide the process of rectifying such behaviour. When the temporary nature of the problem is pointed out by the veterinary advisor, there can be a reduction in the number of affected cats being euthanized (Overall *et al.*, 2005).

Displacement activities

Many activities of domestic cats are displacement activities taking the role of natural hunting. With the latter being frustrated in domestic circumstances, alternative functions become substituted. Many of these are contained in episodes of play, but some take the form of outbursts of kinesis that are directed away from the core area of the cat's domestic occupancy. Efforts to climb curtains are another example. With pair-bonded cats, mutual chasing can appear as aggression, but is more likely to be the discharging of energy intended for hunting behaviour. For motivational diffusion, short episodes of vigorous activity are effective in the cat. A pair has the advantage of using each other as targets for such activity.

A common form of displacement action is when a cat has been stimulated to action but is frustrated by inability to perform. In some cases a cat that has been provoked by another cat, but is unable to reach it, may turn its aggressive mood on the caregiver. A form of displacement has been termed 'feline redirected aggression' by Lachman and Mickadeit (2002). This occurs typically when a new intruder (cat or otherwise) trespasses on the resident cat's immediate territory. The result is that the affected cat can become defensive, agitated or aggressive about the new intruder and turn its territorial response aggressively on another available cat, or even on the caregiver.

All displacement activity can be redirected into playful action by human intervention or, in the long term, by the provision of a more complex environment with cat furnishings such as long shelving, hiding facilities, tunnels, ladders, boxes, etc. Cat furnishings are more needed in multiple-cat households, in order to provide variety in situations that can absorb each animal's interest in favoured placement. Meeting a cat's interest with varied structures helps it to settle into its environment. Territory is a major focus of feline interest, on a par with hunting motivation; the two are primary, cerebral factors in feline life and can give rise to overflow of activity.

Cannibalism

In some species of wild cats it is common for males to cannibalize young kittens or cubs that are not their own issue (Sunquist and Sunquist, 2002). Such cannibalism can occur occasionally in the domestic species also but would be regarded as an abnormal action while housed. For example, this can occur in the circumstances of semi-feral groups of cats on farms, as reported by Macdonald *et al.* (1987). In such a situation the sociology of the group would resemble the wild state, but instances of feline infanticide can also occur in house-kept cats, when an alien tomcat might encounter recently born kittens.

Of less common occurrence is the killing and eating of newborn domestic kittens by their mothers. When such cases do occur, it is noticeable that the offending queen cat may not have prepared a den for the birth or occupied some secluded place by herself when parturition started. With such abnormal prepartum behaviour, the affected queen tends to wander from place to place, giving birth in different locations and may eat each kitten immediately it is born. In this type of case, the obviously confused animal may keep one dead kitten in her mouth and continue to wander about, while holding it in her mouth. No kittens in the litter survive in these cases and the affected queen returns to normal behaviour by the following day. The cause may be some neurohormonal abnormality, but neutering is advised soon afterwards.

Wool-sucking

It is believed that cats with the abnormal behaviour of wool-sucking may have been prematurely weaned from their mothers and that this leaves them with a residual need to find comfort by sucking on palatable material (Fig. 11.2).

Cats with this behavioural condition often choose a variety of materials or garments to chew or suck, destroying them in the process. To break this stereotyped habit, the caregiver has to intervene with an alternative activity, such as active play, or otherwise direct the animal's interest to an alternative focus of attention, such as a favourite toy. It is understood that equivalent forms of stereotyped

Fig. 11.2. Wool-sucking; note expression.

behaviour in various other species release dopamine, which gives an experience of pleasure (Broom and Fraser, 2007), so that the habit is effectively an addiction and may therefore be difficult to break easily. Giving the animal greater freedom of action, such as outdoor release, can help to provide it with alternative interests that can change the behaviour. Outdoor activity can, of course, be provided safely by walking the cat on a leash, after it has been trained to accept the harness. The training procedure for exercising a cat on a leash is detailed by Edney and Bessant (2001) and Bessant (2004b).

Possessive blocking behaviour

Occasionally, the pet cat intrudes into the activity of its keeper by placing itself between the keeper and the item with which the person is occupied. For example, the cat may lie on top of a book or paper which the person is reading; it then remains in that position, obstructing the person's sight of the item. Again, it may lie on the keys of a typewriter or computer keyboard on which the person is working, with the same blocking result. The cat's action virtually puts it in possession of the item that the person is attending to. This possessive blocking behaviour (PBB) can go further by the cat anticipating a person's habit and blocking it, such as in lying on top of a newspaper that has been laid down for later reading. Similarly, the cat may, in anticipation, spread itself across clothing that the person is about to wear.

Another mode of PBB is in the form of the cat obstructing the feet of a person attempting to walk. Some individual cats perform this obstruction very regularly as, for example, when the person is about to leave the house or to walk downstairs. The potential danger in the latter case is obvious. Some cats with the blocking habit usually perform it in one mode only. The place being blocked may be on the floor, bed, chair, table or a step on a stairway, and the cat usually assumes a firm recumbent position on the item.

The cat's motivation in this behaviour can be assumed to be an effort to direct attention to itself. If the cat is immediately lifted from an obstructive position in PBB and held comfortably for a while, it usually purrs strongly and this would seem to confirm that it had been self-seeking for affectionate attention from the person concerned. Such a motive could be related to the feline ego, which is a strong factor in cat behaviour.

The Epigenesis of Sociability

It seems that a significant epigenetic event takes place when the kitten experiences adequate social handling in its second month of life. At this critical stage, the experience of handling apparently activates a genetic factor for sociability. This factor is probably a gene. Failure of genetic activation in the critical period evidently leaves this gene permanently inactive in those cats for which such activation is dependent on experiential induction. With this supposition, an ethological postulate is offered here as follows: Without epigenetic activation of the social factor at their kitten stage, the majority of mature, non-pedigreed cats are left with reduced sociability, relative to situations of human association. With this epigenetic failure, the cat, as a companion animal, is left to learn from experience the successful modes of behaviour at the interface of human contact and association.

Many adult cats in caring custody progressively acquire some degree of social competence by learning, but others (such as feral cats) never achieve any social competence in situations of loose human association. The latter is made very clear in the numerous case histories of Canfield *et al.* (2011). Feral-born cats cannot experience early social handling and are therefore ESH negative. Generation after generation, this is a permanent state in them and one that carries antisocial aggressiveness relative to human contact. However, a young kitten from any one of these cats can become a perfect pet if taken up at an early age, handled humanely and socialized for adoption.

In some domestically reared kittens, there may be degrees of ESH deficiency as a result of being left comparatively untouched with their mothers for extended periods, without being handled to an effective level (whatever this may be, if there is a quantitative requirement). With that possibility, degrees of ESH deficiency can be assumed in many housecats that have poorly developed social intimacy with their caregivers. The key expression of full sociability in cats is the habit of lap-sitting that is so evident among the purebred breeds, as described previously. The activation of a gene for sociability is very clear in these cats.

In the recently formed pedigreed breeds of cats, it is possible that the selection of foundation stock involved the deliberate choosing of animals with a natural capacity for sociability. In the general cat population, rare individuals with very sociable temperaments do occur, without a clear history of early social handling. Managed selection for this trait may have given the pedigreed breeds a sociable genetic factor that became activated in natural epigenesis. On the other hand, the sociability in this population may be simply the result of intense care during kittenhood. Breeding is typically random in non-pedigreed cats and the gene for sociability does not appear to be expressed in their progeny. Clearly, however, it can become expressed by environmental induction, namely through the experience of ESH.

Among the broad population of kittens from non-pedigreed housecats, sociability may not be immediately apparent since the ESH effect has latency insofar as it is not fully apparent until the kitten has been separated from its mother, adopted as a companion animal and become behaviourally mature. Sociability in cats can go a long way in providing them with good welfare prospects. Sociability can be considered as a vital property in a cat that is to have a satisfactory life as a companion animal. The behavioural epigenesis of the kitten has been given special attention in Chapter 4 since this young animal's character can be seen to have norms in its stages of development and these can be assessed by regular inspection to determine normality.

Social negativity

Various forms of anomalous behaviour in domestic cats have the common feature of dissociation from the caregiver. With this condition, the cat's relationship with people always appears negative. The animal's temperament may, or may not, not be one of aversiveness to the caregiver, or others, but a close relationship with people is not evident. Active forms of dissociation can include any or all of the following items of behaviour:

- avoidance of being caught;
- struggling to escape when lifted;
- refusal to approach the caregiver when summoned affectionately;
- tendency to hide frequently;
- frequent Jekyll and Hyde reactions; and
- great difficulty in restraining during grooming or medication.

Together, these features of behaviour, if consistently shown, constitute a syndrome of social negativity that does not disappear with long-term care. Some amelioration of this characteristic, however, does take place with habituation to its domestic environment and the cat can become a satisfactory household animal, when dealt with appropriately. The likely cause of this state is absence of handling during the sensitive period. This state contrasts with the typical behaviour of cats that suffer SAS.

The book *My Cat's Life* (Canfield *et al.*, 2011) presents 101 case histories of cats showing individual temperaments, characteristics and personalities that identified these animals as they coped with a wide range of circumstances. Those cats that were evidently strays or ferals responded positively to supportive care offered to them in the form of food or shelter; these cats eventually became fairly friendly over long periods of time. In no instance, however, did any such cat allow itself to be handled. They remained permanent examples of social negativity that is clearly a feline behavioural anomaly.

Depression

Symptoms and signs of many clinical conditions in domestic animals often include depression (Broom and Fraser, 2007). Animal depression is defined by Blood and Studdert (2000) as '... decreased interest in surroundings ... decreased response to external stimuli'. The depressed animal's activity is likely to be derived from aversive stimulation rather than through spontaneous relationships with the environment. In depression there is a marked depletion of the behavioural repertoire characteristic of the normal animal. The principal features of collateral social behaviour become significantly diminished.

Loss of maintenance homeostasis appears to be an essential criterion of the general aspect of animal illness referred to in clinical terms as depression. The established concept of adjunctive depression in cat illness recognizes the behaviour of the animal as globally changed rather than regionally modified. The main significant measure is in the reduced behavioural frequency of maintenance activities, such as feeding, body care and inquisitiveness.

Depression in a cat requires prompt clinical attention. Diarrhoea and notable inactivity are signs of the emergence of this state. The factors responsible include the malaise attendant on the basic illness and dehydration from the loss of vital fluid and electrolytes. The feline conservation of body fluid is so critically balanced that a major intestinal flux can induce a moribund state. Immediate restoration of fluid and electrolytes is necessary and should be maintained by regular, subcutaneous infusions with a Ringer's saline solution.

Training

With their burden of ancient genes, domestic cats often have motives that conflict with the desires of someone training them. Nevertheless, cats are quite good at learning from their own experiences in the domestic situation. They learn the positions of comfortable places and the significance of customary noises in the environment, such as the sound of the door opening to the outside. They quickly learn the affairs of feeding – even recognizing the sound of the can opener and its meaning to them, if that is the usual prelude to food being given. The clues to training them are there: namely, they learn what they want to learn. For example, they readily learn the cues to feeding and to possible freedom, which are matters of importance to them.

If the cat has ever been disciplined at the hands of its keeper, it is not likely to come promptly when called by that person, unless to receive food. It will not be likely to come when called simply to please its keeper (such as dogs do). The cat is not a 'people-pleaser', by nature. The cat is essentially a self-pleaser, as a result of its genotype that relates back to the state of 'self-living' that is its heritage. That lifestyle is very evident in the accounts of the wild cats with which the domestic cat is closely linked, behaviourally. Evidently, the housecat usually does what it wants, and its principal want is food, even to the point of always being eager to steal food although well fed. This is its weakness, and that can be exploited in training.

Lachman and Mickadeit (2002) describe a method of training for the house-soiling cat that they term S-W-R-R, for startle, wait, redirect and reward. In this procedure the trainer must first catch the cat preparing for the act of soiling away from the litter box. The cat is not punished in any way but is startled with the trainer suddenly making a loud noise by shouting and clapping the hands. The cat runs away. Following a wait of several minutes, the cat is picked up and placed in the litter box. After the cat has excreted it is rewarded with praise or a treat of food. Other forms of startling non-aversively can be used, such as a jet of water directed on the animal from a water pistol. If the cat is found in the process of actively excreting the startle procedure is not enacted. This training method can take up to 16 weeks to be fully effective. Lachman and Mickadeit are professionals in such work and have found, in their experience, that the majority of cases of house soiling by cats involve increased anxiety experienced by the cat plus a reduction in owner–cat interaction (with the owner's anxiety over the problem rebounding on the cat).

The basis of effective training lies in recognizing that the cat responds to positive reinforcement, not to negative disciplinary training. An essentially simple training method has developed using the principle of conditioned reflexes. It is termed the 'clicker' method. Essentially, a hand-held clicker is sounded to coincide with the delivery of a small item of desirable food. The cat is only given the food as a reward for some behaviour that the trainer desires from the cat. The clicking sound is made very close in time to the reward. In time, the cat hears the clicker as a rewarding signal that it appreciates. By this method, the trainer gains control over the cat in specific situations such as remaining in place long enough for the trainer's purposes (Krieger, 2012).

Training by this method has quite lasting effects, but it requires patience and consistency of method by the training person, including precise timing with the clicker. It is claimed that some difficult situations can be brought under control by this method. One notorious difficulty is preventing the cat from darting to the outside door when it is opened, if the cat is a so-called 'door-darter'. It is claimed that by proper clicker training, the cat can be made to 'stay' when the door is being opened. Cats that charge at an open door in order to escape constitute a behaviour problem for which training is needed (Hotchner, 2007). This cat behaviour expert suggests that owners entering the house should ignore the cat for a period before paying attention to it. The objective is not to make entry an occasion of stimulation for the cat.

With cats that are door-darters, training is difficult. It is likely that the home environment needs to be enriched with cat furnishings such as shelved resting places, hiding places, a scratching post, toys and material diversions that attract the cat. Such items are now plentiful on the market. The single, best diversion is an additional companion cat. A pair of bonded cats gives a greatly improved quality of life for both. Exceptions do occur, however, as for example when a single cat has an inbred disposition that makes it aversive to any other cat, or a competitive pet such as a dog. Evidently, cats have to live with the dispositions that they are born with, particularly if these are polarized dispositions. Training, in general, may have only moderate effectiveness on such individuals.

12 Health Monitoring

A major feature of welfare for a domestic cat is the monitoring of its health by regular appraisal (Pitcairn and Pitcairn, 2005). By attention to its physical exterior, a reliable impression can be gained regarding the cat's general condition. The contact of thorough grooming will detect any physical alteration in the animal's bodily state, or in its behaviour while being groomed. Even frequent stroking and handling is a method of monitoring the cat's health status. Apart from checking for any change in the subject's structure or behaviour, it is an advantage to have knowledge of features that have clinical significance, such as parasitism or viral infection (Edney, 1992; Carlson and Griffin, 1995; Siegal, 1997).

A suspicion of any clinical state can allow early attention to the specific disorder. In the clinical diseases of the cat, early veterinary diagnosis and intervention can improve the prognosis. Even with a fatal disease, its early recognition can allow for palliative care and preparedness for the inevitable outcome. An awareness of the most common health issues among cats is therefore a welfare need. This chapter addresses the need for ongoing monitoring of the cat's health. Of special importance is the early recognition of pain (Table 2.2).

Integumental Care

The feline integument includes the skin, hair, ear flaps, paws, tail and claws. These constitute the entire exterior of the animal. As creatures of great curiosity and investigation, their exteriors are exposed to a variety of contacts. Some of these may contaminate the cat's coat and frustrate its hygienic motivations. The cat spends more time licking to clean itself than any other animal and some of this work generates free dander on the coat. Their strong motivation for skin care is also addressed in acts of biting, scratching, rolling and rubbing. In pairs, they often spend time grooming each other. All these items of behaviour indicate that cats have both individual and communal genetic prescriptions for intense grooming. In turn this shows the relative importance of coat care in the well-being of the cat. In domestication, the implicit contract between animal and owner is appropriate care, including integumental care, which is a fundamental item in welfare.

Grooming

Grooming is a key element in the care of a cat (Rubin, 1995). The relationship between good grooming and monitoring feline well-being is close. General grooming includes brushing the body and tail, wiping the face and ear flaps, combing below the tail and wherever the hair may be tangled, smoothing the animal's surface along the direction of the lie of hair. Hair mats should be dealt with before they form thick clumps. The latter become tightened to the skin so securely that the cat will require close trimming under general anaesthesia. Small mats can be cut into at right angles to the body of the cat without touching the skin. Repeated cuttings into the clump break it up sufficiently to be combed out with a coarse-toothed comb.

The session of integumental care should not be disturbing to the cat. If the animal shows serious aversion to this attention at first it should be stopped and the cat immediately given a reward. Actually, routine grooming on a short-haired cat can be completed in one minute. Further sessions can be increasingly detailed within the cat's tolerance until the regimen eventually becomes acceptable to the animal. With short-haired cats frequent grooming sessions are not as necessary as they are with long-haired cats; the latter benefit from daily grooming. While shedding is taking place the circumstances are different and frequent grooming is then necessary.

The caretaking objectives in grooming are hygienic, physiological and inspective – in addition to the provision of some sensory satisfaction to the cat. During the grooming process the practical welfare objectives are: massage of the animal; stimulation of cutaneous circulation; removal of skin debris, loose hairs and skin secretions (in the form of dander); and monitoring the integument for any skin damage, infection or parasitic infestation.

Claw and ear care

Trimming the claws is a job for two people, including someone with an understanding the anatomy of the claw. Only the bare tip should be removed if it is considered that some claw trimming is necessary. Even then, only the forepaws require attention under usual circumstances. When the ears are cleaned it is not necessary to go into the deeper part of the ear. It is sufficient, for purposes of ordinary cleanliness, to wipe the inner aspect of the pinnae with a clean, moist cloth or a damp sponge before a further wiping with dry towelling material.

Templar patches

Every feline has an area of relatively hairless skin on each side of the head between the forehead and the ear. This area of the scalp corresponds with the skull's temple. The skin here is rich in apocrine glands and has a secretory function that is probably pheromonal in nature. When the hair across the forehead is lying flat, this patch of skin is partially obscured, but it becomes completely exposed when the hair on the temples becomes erect. Such exposure occurs at various times, such as when the cat's head is in direct sunshine. At such a time, this patch has the shape of a wide isosceles triangle having its peak above the inner canthus of the eye and its base at the opening to the ear on the same side of the head.

These templar patches of skin are the focus of much of the healthy cat's grooming activities. Typically, in the course of self-grooming, the cat licks the hair over the carpus and metacarpus of one paw and the proceeds to wipe the inner aspect of that paw over the temple on the same side. This is done with some vigour and is repeated several times as an action pattern (Fig. 12.1a and b).

The other side of the head receives equivalent grooming. Cleaning this area evidently has some special importance in the cat's self-maintenance. It is likely that the patch exudes a scent and is better able to radiate that scent by virtue of being scarcely covered with hair. Rubbing the side of the head against objects would also place the cat's scent firmly on them. The presumed scent would be the product of each individual cat and would therefore supply the animal's identity. Such a scent is presumed on the basis of circumstantial evidence, but is beyond detection by the human sense of smell, as are most animal pheromones (Wyatt, 2003).

Flea infestation

At some time, the majority of household cats become hosts to the cat flea (*Ctenocephalides felis*). This bloodsucking, ectoparasitic insect seeks its hosts everywhere. It can live away from the cat for 7 months without ingesting blood. It thrives best in warm and humid environments. In temperate or cold climates it is active during the summer. In an indoor environment the flea is active perennially. In its reproduction the flea is prolific and one female can lay hundreds of eggs throughout her living space, depositing them on carpets, floors, the cat's bedding, etc. From egg, to larva, to pupa, to free-living flea takes only 3 weeks.

The flea can detect any nearby cat and, with its spectacular jumping ability, it can infest a cat in an instant. It feeds by puncturing the skin with its pointed mouthparts, using its saliva to facilitate the penetration to a capillary. It then sucks a quantity of blood. This activity is not instantly detected by the host, but when the saliva is absorbed by the skin surrounding the puncture a localized inflammatory reaction quickly occurs, causing an intense itching at the site. The cat shows the symptoms of infestation in its scratching behaviour; episodes of scratching are frequent and vigorous.

When the flea population increases, the parasites begin to seek additional hosts such as humans and dogs living in the premises. As a rule, the cat flea on a human bites the lower parts of legs, causing hypersensitive local reactions, or sometimes more widespread areas of the skin become affected with intense pruritus, calling for localized treatment with an anti-inflammatory agent.

A heavily infested cat is likely to develop hypersensitivity to the flea's salivary antigen and the skin lesions become more severe than singular bites. In such cases the cat can acquire 'scabby cat disease' with miliary dermatitis on the back that produces papules and crusts. The cat's scratching behaviour intensifies. Flea infestation can be confirmed by brushing debris from the cat's coat on to wet

(a)

(b)

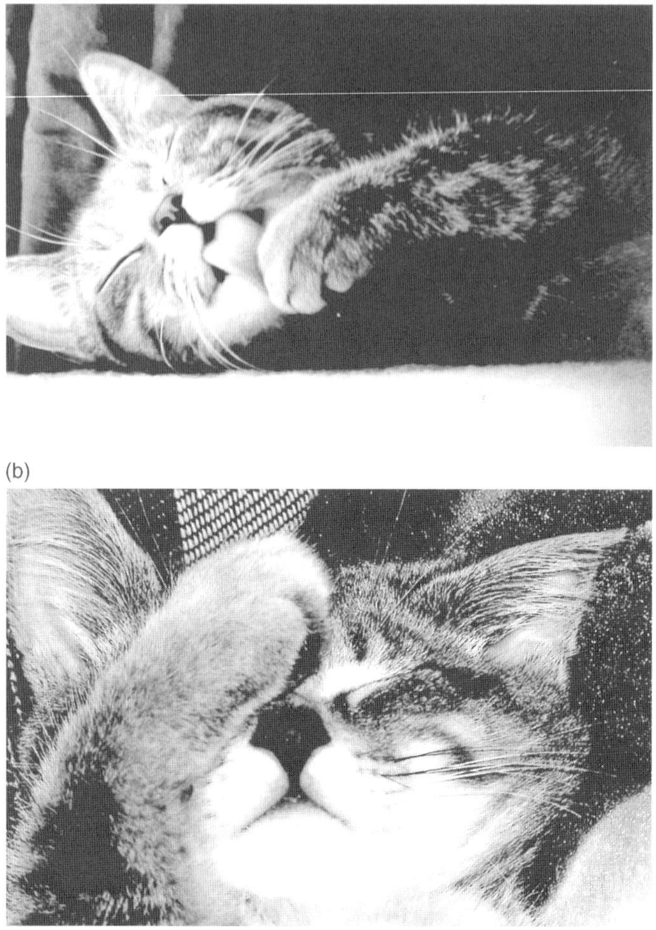

Fig. 12.1. (a) Wetting the paw to sponge the temple; (b) kitten wiping temple.

white paper, when 'flea dirt', composed largely of cat blood excreted by fleas, turns to small reddish-brown blots.

One reason for presenting this condition in some detail is that it constitutes a common problem, countering the cat's well-being. Another reason is that cat flea infestation is a major welfare problem in cat care. Cat owners, affected themselves perhaps, become intolerant of this infestation in the home. They deplore the cat's disturbed life and are often frustrated by their own inability to eradicate the parasite completely. As a result, some owners try to resolve the problem by getting rid of the cat and getting the home disinfected by fumigation.

A plan, in principle, for the control of cat flea infestation should be available to cat managers in general. The condition requires two forms of attack to be commenced simultaneously and as quickly as possible. One of these is aimed at the domestic environment and the other concerns treatment of the cat.

Treatment of domestic environment

An insecticidal powder, such as one containing 5% carbaryl with a trace of pyrethrins, can be applied to rugs, carpets, mats, cushions, soft furnishings, the cat's bedding and its common resting places. The powder should be worked into the materials of these items. Any lightweight coverings contacted by the cat can be washed at the same time. Frequent and thorough vacuum cleaning of floors is necessary and some of the insecticide powder should be vacuumed up occasionally to mix with the contents of the bag.

Treatment of cat

The modern method of treatment for cats over 8 weeks of age is by applying a spot-on preparation once per month for 2 months, if deemed necessary. This involves putting a minute quantity of the therapeutic liquid on to the cat's skin between its shoulder blades where it cannot be licked off. This may require an assistant to hold the cat and part the hair properly so that the agent is put directly on to the skin. It is absorbed progressively from there and circulates in the bloodstream. Various products with similar action are available, such as imidacloprid, fipronil and selamectin. These are marketed under trade names. Another highly effective product for flea control in cats is lufenuron, which can be given by injection at 6-monthly intervals. A flea infestation in cats can be easily combated by such methods with veterinary advice.

Mite infestation

All of the big wild cats have ears of moderate size with rounded tops while, with the exception of the manul, the smaller wild cats and the domestic one have ears extending prominently and firmly up or out from the head. The feline ear typically has a tight covering of hairs stretching from the inner edge towards the outer edge of each pinna. This ear design permits both ventilation and screening of the ear and its outer canal, making it a healthier structure than is the case with many canine ears. Nevertheless, the domestic cat's ears can occasionally become affected with some inflammatory condition, such as mite infestation.

Mites can infect cats that have low tolerance because of age, i.e. very young or old, or poor welfare. Mites can concentrate on the ears or on the skin around the face. The mange mites, i.e. sarcoptic or demodectic, invade the skin or hair follicles causing lesions where they locate. Demodectic mites live in hair follicles on the cat's ears, chin and around the eyes, causing bald patches that itch. Sarcoptic mites burrow under the skin, feeding off superficial blood vessels and creating scab lesions that form on the long narrow tracts of vessels.

Ear irritation may be caused by mites. The result is an accumulation of offensive smelling dark brown wax, plus some swelling of the ear canal and considerable itching. The itch causes the cat to scratch at the ears and tilt its head from side to side repeatedly. Scratching can result in abraded skin around the area of the affected ear. Treatment is basic cleansing of the ear's outer vertical canal with twists of cotton wool moistened with warm water to swab out the excess of contaminated wax, followed with application of an acaricide solution. The hindfeet can be cleansed with a solution of hydrogen peroxide. To deal with re-infestation, repetitions of treatment must be done on days 9 and 18.

Parasite control

The internal parasites that can diminish a cat's well-being are chiefly roundworms, tapeworms and hookworms. Formerly internal parasitism treatments involved medication by mouth, with its attendant difficulties. This often discouraged cat owners from carrying out anthelmintic treatments as often as needed. A simple, topical medication is now available for the treatment and control of roundworms, tapeworms and hookworms. Its composition is emodepside/praziquantel and is given by applying a small quantity of the liquid to the skin of the cat at the back of its neck, close to the head, in a manner similar to the treatment for fleas with a topical liquid. Examples of these have the trade names Profender and Advantage.

Feline Disease

Viral diseases

The greatest threat to the cat's health and life come from infection by any one of its different viral diseases (Quinn *et al.*, 2000; Gregory, 2004). Three of these diseases can be avoided by vaccination, while the others cannot be controlled by immunization. These six are the main diseases of cats that destroy well-being; they cause extreme suffering and often death. These viral diseases are:

- Feline panleukopenia virus (FPV) – vaccine available.
- Feline herpes virus (FHV) – vaccine available.
- Feline calicivirus (FCV) – vaccine available.
- Feline leukaemia virus (FELV) – no vaccine.
- Feline infectious peritonitis virus (FIPV) – no vaccine.
- Feline immunodeficiency virus (FIV) – no vaccine.

Young cats that are newly acquired must be kept under quarantine until immunization has been completed for the first three diseases listed. The best protection against the remaining three diseases can only be provided by lifelong quarantine as an indoor cat. Regular outdoor exposure is very likely to result in the cat contacting these viruses. Some cats may have natural resistance to these infections, but these would be few and unidentified. These infections are also transmissible to most wild cats.

Feline panleukopenia virus (FPV)

This deadly viral disease of cats goes under a confusing variety of names including 'feline infectious enteritis' and 'feline distemper'. The disease, with an incubation period of 2 to 10 days, typically has sudden onset with arrest of appetite, vomiting, diarrhoea, pyrexia and depression. Severe leukopenia develops, with hypothermia preceding death within 1 week, in most cases. In adults, the mortality rate varies between 60 and 90%. Kittens may die in the first day of illness. Survivors continue to shed the virus for 6 weeks after recovery, with the result that most outdoor cats are very likely to be exposed to the infection by direct contact at some time (Sigmund, 1979; Kahn and Line, 2010).

Contaminated premises can be disinfected by washing with a dilution of bleach. Indoor containment is the practical means of control, but the highly infectious nature of the disease demands the protection of vaccination. The vaccine should not be administered to kittens under the age of 4 weeks or to adults that are sick in any way. Kittens should be first vaccinated at about 2 months, then at 3 months and possibly at 4 months of age. The immunization should be completed at least 2 weeks before any possible exposure.

Treatment of this disease is mainly directed at rehydration with Ringer's lactate solution and parenteral antibiotic therapy plus supportive nursing, such as warmth and the offer of easily digested food. Cases that live for more than 7 days are usually capable of surviving and such cats should then be quarantined for 6 weeks.

Feline herpes virus (FHV)

The disease of cats caused by the herpes virus has also been called 'feline viral rhinotracheitis' and 'cat flu'. Many of the common upper respiratory diseases in cats are the result of FHV infection. A dual infection with FCV can occur. Transmission of FHV is usually by droplet infection or sometimes by fomites in a cattery.

The onset of the disease is characterized by frequent sneezing, rhinitis, conjunctivitis and a temperature up to 40.5°C may develop before subsiding to about 39°C. Excessive salivation may also occur and a serous discharge from the nose and eyes develops.

As the disease progresses, the discharge from the nose first becomes purulent, then the nares become encrusted with dried discharge. The cat shows deep depression and absolute anorexia. Those cats that become most debilitated develop ulcerative stomatitis and a few may acquire ulcerative keratitis.

If there is significant secondary infection, antibiotic therapy would be indicated. Careful attempts should be made to clear the nostrils using a warm, wet sponge to relieve the animal's suffering and to encourage the return of appetite. The use of the sense of smell is a factor in the cat's ingestive method.

The disease can occur in any age group, but the incidence is much higher in kittens and older cats. The course of the illness may be as short as 1 week or as long as 6 weeks. Most cases survive, the prognosis being good for a strong mature cat, but less certain for very a young kitten or an aged cat. FHV vaccination against this disease should always be done to the maturing kitten to give it lifelong immunity.

Feline calicivirus (FCV)

Like FHV, FCV invades the upper respiratory tract causing, in this case, oral ulceration, gingivitis, stomatitis and upper respiratory disease. The characteristic lesion is ulceration of the oral mucosa after small vesicles on the tongue or hard palate have ruptured. The virus has different strains. Some strains can cause pulmonary oedema and interstitial pneumonia; other strains can cause acute, but transient arthritis in the limb joints of young kittens creating the condition of the 'limping syndrome'. The condition of lameness changes from leg to leg, but soon resolves without treatment.

Commonly, with the oral condition, there is a short-lasting fever. The mouth lesions tend to heal rapidly and the duration of the other symptoms, such as depression and inappetence, is limited to 7 to 10 days. Other cases may be more serious with pneumonia developing.

Diagnosis is based on all the clinical findings, notably sneezing, rhinitis, conjunctivitis, lacrimation, salivation and the oral ulcers. Treatment is in the form of supportive nursing by keeping the nose clear of accumulated excretion, for example with nose drops containing an ephedrine solution. Spontaneous recovery is usual. In cases where a secondary infection has occurred, it could be necessary to provide antibiotic therapy. Vaccination programmes, using different vaccines, can give substantial protection against FCV and all domestic cats should be vaccinated against FCV for their well-being. In addition, precautions can be included in cat management by preventing any overcrowding of cats and by keeping a cat from outdoor exposure to other possibly infected cats.

Feline leukaemia virus (FELV)

Another deadly virus affecting cats is FELV, which is highly contagious and is often continuously present in the saliva of infected cats. Transmission is by cat-to-cat licking and biting. The disease caused by this virus may also be called 'feline lymphoma', or 'lymphosarcoma'. The neoplastic condition of lymphosarcoma is the most common form of cancer in cats. The lesions occur in the thorax or, less often, in the abdomen. The disease has its highest incidence in multi-cat households.

Some cats that become exposed to the infection develop a natural immunity to it and remain free of the disease, but most acquire the infection as a permanent condition and become chronically ill. Typically the diseased cat suffers a long-term wasting disease with anaemia, loss of appetite and general debility. A lymphosarcoma in the chest causes difficult breathing, while a diffuse or a discrete tumour in the abdomen will cause enteritis and malabsorption of food with cachexia. If the kidney becomes affected, uraemia develops. Involvement of the liver causes jaundice to become apparent. Infection can attack the gravid uterus and abortion may occur. Otherwise the kittens are born with the infection for life.

In addition to causing its form of cancer, namely lymphosarcoma, FELV infection can cause other clinical conditions, notably anaemia and immunosuppression. The latter, of course, allows a range of opportunistic pathogens to infect the animal. Such pathogens may include any of the other virulent feline viruses, such as FIPV.

A laboratory test for FELV can detect infection and carriers, but some cats with a positive test may remain healthy and should not be condemned on the test result alone. Those with negative test results should be vaccinated. The vaccination involves two initial injections and a third later followed by an occasional annual one. Disinfection of premises where a diseased cat has been living may not be required to be intense since the virus does not remain alive in the physical environment.

Feline infectious peritonitis virus (FIPV)

With a mortality rate of almost 100%, feline infectious peritonitis is another deadly contagious disease caused by viral infection of both domestic and wild cats worldwide. It affects all ages and both sexes, but the incidence is highest among very young and very old cats. Stress can also contribute to vulnerability (Peterson *et al.*, 1991). Infection is acquired by oral contact with contaminated excretions left in the environment. The disease has a wet form and a dry form. Many cases also carry FELV (Kahn and Line, 2010).

The disease is often first manifested in a prodromal range of pathological conditions such as pyrexia of unknown aetiology, conjunctivitis, respiratory tract infection, or diarrhoea. Additional clinical conditions include anorexia, loss of weight, anaemia, persistent diarrhoea, lethargy and progressive debilitation. A fever that does not respond to antibiotic therapy is another finding in the prodromal period, which may last for weeks before the disease proper appears.

The most common age groups to suffer from feline infectious peritonitis are those that are not yet fully grown, the adolescent cats. This fatal disease is caused by a mutated corona virus and transmitted by contact with an infected cat. The principal symptom is severe ascites. The volume of fluid in the abdominal cavity accumulates progressively, making the animal increasingly pot-bellied in appearance. Other signs of feline infectious peritonitis include haemorrhages into the eye and jaundice, which cause the cat's facial appearance to change very noticeably. When clinical signs have developed, death ensues. Older cats can carry the infection for lengthy periods without showing clinical signs. Diagnosis can be difficult in many cases, but a method of biopsy could confirm a diagnosis. Otherwise diagnosis is dependent on the case history plus the clinical findings.

In the wet form of feline infectious peritonitis there is fluid effusion in the peritoneum, the abdominal cavity and sometimes the pericardium. In the dry form granulomatous lesions develop in any organ of the body including the central nervous system and occasionally the eye also. The disease signs at that point are determined by the parts of the body that lose their functions. With organ failure the cat becomes moribund and dies.

Certain diagnosis is difficult, but a presumptive diagnosis can be made on the prodromal history and the finding of fluid in the peritoneal cavity or thorax. The disease is incurable and no effective treatment is known. All that can be done is to provide comfort to the animal until the arrival of the moribund state, when euthanasia is necessary (Houpt *et al.*, 1996).

Feline immunodeficiency virus (FIV)

The disease caused by FIV is sometimes referred to as 'cat AIDS'. Insofar as it is the ultimate immunodeficiency that creates conditions for opportunistic infections, there is similarity. The condition is endemic in the feline population around the world, including domestic, feral and wild cats. The virus is shed in saliva and transmission is by biting. As a result, more male cats than females become infected through fighting, since biting is the main method in fights. This is how the inoculation of the virus occurs.

After infection has happened, a transient fever develops. This causes some evidence of temporary illness from which most cats soon recover. The infected animal may then appear healthy for some time, even for several years. When immunodeficiency develops eventually, the terminal phase of the disease begins. The animal is now defenceless against a range of opportunistic pathogens that create chronic illnesses.

The body regions most vulnerable to such secondary infections are the respiratory tract, the gastrointestinal tract (including the mouth), the urinary tract and the skin. Neoplasias develop, as well as neurological disorders from disease of the cerebral cortex. The latter state gives rise to various behavioural abnormalities, psychomotor disorders, dementia and convulsions. Euthanasia is the only course open, since there is no cure and all the possible symptoms add up to suffering that is unnecessary.

Feline urological syndrome

A very common disorder in cats is the condition termed 'feline urological syndrome' (FUS) that can occur in both sexes with equal frequency (up to 10%), but affects males more acutely by causing obstruction to the outflow of urine. In the female, the more common consequence is inflammation of the bladder and urethra, which is of no less significance with regard to the degree of illness and loss of well-being. It is also termed 'feline urolithiasis', since the cause of the condition is stony substance in the urinary tract. The stones responsible are usually composed of magnesium-ammonium-phosphate (struvite). Less common uroliths in cats include calcium oxalate and ammonium urate.

Most cases occur in cats kept indoors, usually in their younger years. The explanation for this incidence is that the relative lack of physical activity from confinement and neutering in this group is associated with minimal water intake and resultant concentration of urine. This concentration allows the mineral content of urine to form crystals and these adhere to become stones. The mineral most at fault is magnesium.

Diets with high fibre content can result in the faecal matter holding more water and therefore reducing urinary volume. Some canned cat food with reduced calorie content induces the cat to consume more solid matter with a disparity in corresponding fluid. This results in some diversion of fluid to the gut instead of the kidneys. On the other hand, some dry cat foods contain lower fibre and magnesium than others and this can help in the prevention of FUS. The actual level of magnesium per volume is the critical factor, which obviously favours the lowest possible amount of magnesium.

With crystalline matter in the urinary tract of the male cat, there is greater opportunity for it to become lodged in the urethra, it being both narrower and longer than in the female tract. In either sex, a stony substance can abrade the mucosa of the bladder, causing inflammation, irritation and pain.

Treatment, in principle, consists of fluid therapy and relief of the obstruction. The latter can often be done by flushing the urethra with fluid consisting of equal parts of isotonic solution and an aqueous lubricant. Occasionally there is a need for cystocentesis, tapping the bladder via an abdominal mid-line puncture, if it is markedly distended.

This would be done before any attempt at hydropropulsion against the obstruction.

Prevention of the condition is effected by feeding a diet containing less than 20% magnesium per 418.4 kJ (100 kcal) metabolizable energy. It is also important to prevent the urine from becoming even slightly alkaline. The cat's urine must have a pH under 6.4. A diet for this purpose has been given by Merck & Co., Inc. (Fraser, 1991: 890) as shown in Table 12.1.

Feline lower urinary tract disorder

Feline lower urinary tract disorder (FLUTD) is a painful problem for any cat affected. Predisposing factors include neutering, obesity, lack of exercise and dry food. The type of cat most commonly affected with FLUTD is of the Persian type kept indoors and aged about 3 to 8 years old. The behavioural signs include frequent visits to the litter tray, straining to urinate, frequently licking the genital area, passing only small amounts of urine per effort, or in extreme cases passing none. Some cases go into a damp area, such as the bath, while attempting to urinate and this is very indicative of FLUTD at a stage requiring immediate veterinary attention.

Diseases of pedigreed cats

Although pedigreed cats are as healthy as unpedigreed cats, they have certain genetic diseases that are uncommon in ordinary housecats. Eight of these are described by Eisen (2012) and are given below.

Table 12.1. Homemade diet for the management and prevention of feline urological syndrome. Combine all ingredients and feed 110–230 g daily. Adapted from Anon. (1984) *Veterinary Medicine/Small Animal Clinician* 79, 334.

Ingredient	Quantity
Ground beef, braised lightly and fat retained	450 g
Uncooked liver	110 g
Cooked rice	230 g
Cooking oil	5 ml
Antacid pills, ground	5 g
Water, added during cooking	60–90 ml

1. *Polycystic kidney disease*. This condition shows up in adult life, although congenital. Mostly Persians are affected. The behavioural signs include loss of appetite, depression, thirst and polyurea.
2. *Hypertrophic cardiomyopathy*. The heart walls become thickened in this condition, leading to cardiac arrest in kittens. The breeds affected include Maine Coon, Norwegian Forest, Ragdoll and Persian. The behavioural indicators include intense drowsiness, thirst and the related increase in urination. Hypertrophic cardiomyopathy can occur in non-pedigreed cats, of course (Boland, 2011). It is the leading cause of sudden death in the wide population of cats.
3. *Osteochondrodysplasia*. This is a problem in the Scottish Fold breed. The behavioural signs include loss of curiosity, loss of interest in play, reduced appetite, increasing lethargy, lameness, stiff gait; all of these leading to paraplegia.
4. *Renal and hepatic amyloidosis*. With an increase in the amyloid content in the two detoxifying and excretory organs, affected cats become progressively ill showing the symptoms of increasing thirst and weight loss. Polyurea accompanies the extra fluid intake from the thirst. The Siamese breed can be affected with this rare condition.
5. *Progressive retinal atrophy*. This condition goes through stages of diminishing sight, leading to a high degree of visual impairment by the age of 2 or 3 years in some cats of the Abyssinian breed.
6. *Brachycephalic problems*. As a result of the fashion in the short-faced breeds, such as the Persian, to have these cats with very short muzzles, difficulties in breathing can develop. The resulting stenosis of the nares can lead to inefficient respiration. The sinuses can be involved in the disorder and the lacrimal flow may not be adequately drained. Eye lesions can result.
7. *Haemophilia*. A failure in the clotting of extravasated blood is an occasional problem in some breeds including British Shorthair, Devon Rex and Siamese. The condition is carried genetically by females in the affected strains, but the clinical state develops in their male offspring. Kittens can be affected at parturition and be stillborn or die soon after birth. Older males have uncontrolled bleeding at surgery.
8. *Hip dysplasia*. As a result of the hip joint being abnormal in shape, the femoral head may not articulate easily in the acetabulum. This condition is encountered in the Maine Coon breed, causing difficulty in walking and reluctance to jump.

The breeds affected by the conditions indicated above require the neutering of the affected breeding cats. These states cause suffering in the cats bred with them. Their existence in an individual is inconsistent with a state of well-being.

Feral Cat Health and Welfare

Around the world, many millions of cats of the domestic species now live in a feral state, usually in small groups or larger colonies (Izawa *et al.*, 1982). Many of them become infected with pathogens, diminishing their health. The feral state is not always similar to a wild state, since these cats – while free – remain connected marginally to human presence. Some are in rural areas and farms, but most are adjacent to human activities in places such as factory sites, public parks, hospital grounds and industrial yards (UFAW, 1995). Their existence is a mirror image of their first step towards domestication in the Levant about 5000–7000 years ago. At that time, some had drifted away from desert lands to scavenge around burgeoning human settlements and also hunt the rodents that were doing the same thing. The cats were given a welcome, the rodents were not, and the cats were readily taken into homes and domiciled, as their inherent attractions became evident (Serpell, 1988).

In the modern feral condition cats have simply taken themselves back out of domesticity – not to the Western Desert, however, but to some local retreat that provides access to scraps of food and a degree of communal shelter. In brief, the feral cat has merely returned to its beginnings (Tabor, 1983). This is yet another example of the very persistent genotype and low immutability in this animal, hammered into it in a demanding and protracted evolution.

Most feral cats are born from feral parents, but additions to feral groups come from lost, displaced or abandoned cats. Each cat's state of being feral is not necessarily a result of its intention, but in most cases is the result of misfortune. In fact, feral cats are simply homeless cats that are attuned to an antisocial lifestyle. A large collection of anecdotes on feline lives gives sufficient proof of that (Canfield *et al.*, 2011). These true stories show that free-living domestic cats are true to their species in their sundry personalities.

If there are fertile cats within a feral group, the group size will grow and its population will get out of control with fighting toms and limited food. Such a group, generating noise, odours, faeces and dying kittens, becomes a public problem in its locality. In most cases the complaints from people resident in the area result in efforts to eliminate or control the feral group. In some regions the objection to feral cats is based on the fact that such cats can kill local birds. In Australia, for example, some rare, native birds (of species that never experienced feline predation before, Australia being without any native wild cat) are on the point of extinction as a result of being preyed upon by feral cats.

Formerly it was common for authorities to arrange for an agency to catch and kill such cats, but the welfare movement has succeeded in changing that system of elimination to one of control, by trapping, neutering and returning (TNR) each healthy cat in the feral group. The catching procedure involves feeding the cats on a schedule and placing some feed near a trap, then placing feed further and further into the trap – far enough for a hidden person to pull a cord that will close the trap. Some traps can operate automatically with a trip switch closing the door behind the cat when it enters the trap. While this is an advantage in capturing the animal, it can be a disadvantage since the cat reacts violently to its sudden entrapment and will continue to do so for some time. This can be cruelty if the loaded and uncovered trap is left unattended for a long period.

The emotional stress of such an experience for the animal is sometimes excused with the view that the cat may be euthanized anyway. It is not an uncommon view outside the discipline of Animal Welfare that euthanasia wipes away suffering. While it can terminate suffering, it in no way eliminates any prior suffering, such as being writhed with anguish, which is a negative state of immediate experience for which euthanasia does not give retroactive relief.

Transferring the cat from the trap to a carrying cage is then necessary as soon as possible and this can be a difficult procedure for an inexperienced person. With the cat in a carrying cage it can be promptly taken to the veterinary clinic. It is very desirable to have the cage covered to reduce the stress on the cat. The cat's distress is limited when its vision of its own territory ceases and it can concentrate on its secure container.

At the veterinary clinic each cat receives a full clinical examination and it will be found in many cases that some individual animals have conditions causing significant suffering. Such conditions will include advanced disease, organ failure, chronic debility and loss of essential functions. It would be inhumane to allow continuous suffering in such cats and prompt euthanasia is therefore necessary. Any health disorder that can respond to treatment would receive appropriate attention before being returned. At the conclusion of TNR, for all cats in the group there should be a general state of good health in that population. Follow-up observations are needed to attempt to maintain a satisfactory state in the group. This ideal, however, may be of limited duration. Large groups have the most problems (Kerby and Macdonald, 1988).

With no further breeding, the group can be left to its devices with feeding support from some party in the district. Feral cats are unsuitable for re-homing, since their aggressive and defensive behaviour is chronically established in them. Similarly, they cannot be placed in an established animal house, such as an animal shelter belonging to a humane society, because of possible disease transmission and difficulty in handling. Sometimes they can be accepted on a farm where they can be useful in controlling vermin (Macdonald *et al*., 1987). Some farmers are satisfied to have feral cats and they often feed milk to their cats. Under such better conditions of welfare, the animals are really in a semi-feral state.

When the group members have all been treated with the TNR management system, there is usually social stability in the group; however, there is likely to be continuing loss within the group from road accidents, disease and age. The heart of the colony can be improved for the welfare of the cats by providing water-proofed and bedded boxes as shelters. As a group becomes reduced in size by attrition, it is common for new arrivals to be tolerated and there is therefore a constant need to check for newcomers so that they can receive the TNR management. A record can be kept of group members to monitor the composition of the group or colony and also to log all significant matters concerning the feral population concerned.

A constant threat of infectious disease exists within any feral group. In particular, the viral infections of feline leukaemia and feline immunodeficiency disease haunt feral colonies. It has been estimated that about 10% of feral cats are carriers of these diseases (UFAW, 1995). Checking every cat in a feral group for carriers of these infections is not possible practically in most cases, but the veterinary examination would pick out any individual cat suffering clinically from such diseases. This would allow the exclusion of such cases from returning to the colony. Vaccination and culling diseased cases help to relieve feral colonies from the worst of infections, but many other health problems, such as internal and external parasitism, can prevail in a feral colony.

Without any breeding females in the group, any intact males leave and fighting then becomes rare. The individual cats in the group become friendlier to one another after this procedure (Fig. 12.2).

Fig. 12.2. Group of feral cats assembled at feeding time – friendly but untouchable.

With the noises and odours from toms removed from the group's territory, it becomes more acceptable to local people. TNR is now the recognized system of dealing with feral cat groups of any substantial size. It is currently being practised on a substantial scale in some countries, but it is dependent on volunteers and charitable support for the costs. Volunteers have to be advised of the risks of injuries from scratches and bites in the course of dealing with these cats, since these animals can be aggressive. In addition, there are risks of injury while seeking out cats in derelict buildings. Protective clothing is needed for such work.

Obviously, there is a need for organized arrangements in this system (Kortis, 2004, 2007). Catching each cat requires trapping and transport. A veterinary clinic has to be involved to examine each cat to determine its health status, give treatment for parasites, vaccinate, neuter and perform euthanasia where clinical findings indicate the need for a humane death. After surgery with general anaesthesia, neutered females require a night or two under care before they can be returned to a free state. Every cat treated in this system requires identification and the recommended method is taking a slice off the top of the left ear so that the cat can be recognized by its silhouette without the need of capture. This is done while still anaesthetized and the ears are cleaned before the ear mark is made, to prevent any subsequent scratching which could cause unnecessary bleeding from the cut edge of the ear (Cuffe *et al.*, 1983). This entire protocol is in accord with humane practice.

Very young kittens found in the group can be taken away at about 6 weeks of age for customizing by handling and given a home as a pet. After 2 months such kittens are not amenable to domestication. Any kitten taken from the group would require its health status to be monitored regularly and vaccinated. Such a kitten will be infested with roundworms and fleas, and treatment for such infestations is immediately needed. Since ringworm is occasionally another problem among feral cats, it would be necessary to have the kitten examined under a Wood's lamp for a positive result. Early cases of ringworm in cats can show up positive as a result of the fungus in the hair tips, even without an obvious skin lesion.

The return and release of feral cats, after being neutered and suitably treated, places some responsibility on those who planned and operated the procedure. It is common for feral cats to have the benefit of a voluntary feeder and such a person should be recognized and assisted in this great commitment. Of essential involvement in any TNR scheme is the participation of a veterinary clinic and the costs of surgery and treatment must be included in the budget for the scheme. The continuation of such support for the animals gives them some assurance of a reasonable life. Some get a niche in the community where an excess of human food is available (Fig. 12.3).

At this time, the majority of feral cats do not become involved in a TNR scheme and this

Fig. 12.3. Small group of homeless cats at fish market.

omission is an unfortunate matter in society's current circumstances. Without adequate veterinary attention or the supply of food, many feral cats die eventually of illness or starvation. Their plight represents a major concern in animal welfare.

Housed colonies

While they are not truly feral, large colonies of cats are sometimes kept in houses. These house-kept colonies often have populations of several dozen cats breeding freely with the result that they are co-related and composed of many generations. In such cases it is common for the colony to be held in a secretive state by owners who may be well intentioned but incapable of controlling their cat population that has usually emerged from a few original pets. In addition, it is usual for such owners to maintain their cats without veterinary attention, to disapprove of the surgery of neutering and also to be averse to conventional opinions on animal care. As a result, such cats may not receive any preventive medication or treatment for disease control and without vaccination the colony usually becomes infected with viral disease, many cats becoming ill.

When the increasing colony size and poor health status create a situation well beyond the control of the ownership, the existence of the colony soon becomes known in the community or neighbourhood. The public health authority is notified and promptly takes action to deal with the problem. As a rule, the matter is then reported to the police and the local branch of an animal cruelty prevention organization. These, in turn, call for veterinary involvement. Very quickly, the entire, often complex, problem is placed in veterinary hands with support from the official regulatory body. The veterinary party should follow a protocol of procedure from the start and this can be along the following lines:

- A clear arrangement should be established for the responsibility of professional fees, the manner of caring for healthy cats and for the euthanasia of cats in any state of suffering.
- Authorization to enter the premises and take cats must be determined.
- The involvement of a social worker should be arranged in the likelihood that the owner will be distressed.
- Official approval should be obtained for euthanasia of any cats by the veterinary specialist. It would be desirable to have the owner to agree to this also.
- Arrange with the animal humane society the management of all cats found healthy on examination. If there are no legal proceedings likely against the owner, the return to the home of a small number of surviving cats can be considered, following disinfection of the premises.
- Cats should be taken to the veterinary clinic for examination, assessment and euthanasia, where a state of suffering is determined.
- Prepare a written report of all findings and actions taken for issue to appropriate authorities and for record.
- Be prepared to be an expert witness in court if legal proceedings develop.
- Arrange for follow-up examinations of surviving cats.

Although such cats are familiar with human presence they may not be accustomed to handling, so human safety has to be taken into account when the cats are being caught and examined. If any cats are to be re-homed, their behaviour should be assessed in the course of the veterinary examinations. Any clinical tests considered necessary for such cats can be performed after the initial procedures have been completed.

Although they are not socialized and are negatively disposed to human control, feral cats must be regarded as the same as other housecats. They can have a cautiously friendly attitude to the humans who feed them or give them some unthreatening attention. They can come close to such people, but will not tolerate being touched. Their reaction to any attempt at being handled is fierce withdrawal. This does not make them abnormal, simply non-socialized (ESH deprived). The way that they form close-knit groups resembles the pride formation of lions. They give clear proof that the demands of harsh circumstances for cats create a truce of collective tolerance for survival, notwithstanding their alternative mode of living in a solitary state when their needs of food and comfort, as individuals, are fully met in domiciliary security. Although they have independent habits, feral cats, like housecats, have a need for appropriate welfare.

13 Points of Welfare

A cat owner might ask: 'What is welfare in relation to cats?' Here is a useful question, since there are foggy notions about animal welfare in general. Welfare now has several connotative meanings. Among those who are in support of the ideals of the animal rights movement, it basically means giving the animal good conditions for living in its natural ways and not doing things that are opposed to the animal's interests. The intent is to allow the animal to enjoy a mainly unfettered life. For those addressing animal welfare as a scientifically based subject in need of promotion, it means the determination of the animal's well-being as it copes with the circumstances of its husbandry and use. The main objective is to recommend to animal users and authorities the best possible conditions of husbandry for all livestock, on the farm or in the home.

For those concerned chiefly and realistically with optimal standards in animal care (and in cat care in this instance) it means doing what is best for the animal in three areas, namely: (i) providing for a life of less stress; (ii) facilitating behaviour that is normal; and (iii) maintaining good animal health. The objective is the establishment of well-being. This latter threefold model of welfare is a comprehensive, or holistic, view of animal welfare as a system of practical support and action via ethical animal care. The ethical policy is the pursuit of proper practices in animal use to protect the animal from unnecessary stress, to guard its health and maintain its behavioural soundness as far as is possible in its domesticated circumstances. To put it simply, these ethical principles in animal welfare are to establish a fair deal for the animals with wild ancestry that we have naturalized within our civilization.

As co-inhabitants in our homes, cats add an extra dimension to our own existence. As human living standards have improved greatly in recent decades, so it is time to extend some of that improvement to the cats that are companion animals. By making fine adjustments to conventional housecat management, we can improve the common systems of cat care and give better lives to our pets. Any cat owner with experience can think of modest improvements to a cat's routine existence that will help the cat in some way, or give the animal more comfort and interest, such as playing with objects or partners (Fig. 13.1a and b). Routine features of care can give major returns in cat satisfaction or health.

Twenty cat welfare directives are given here, but imaginative and considerate owners can always establish additional aids to enhance life for their own cats.

Welfare Guidelines

- City cats should be kept indoors in central urban areas and have comfortable, warm quarters.
- All food should be fresh and water always available.
- The food should vary slightly on occasions to aid appetite.
- The litter box should be kept clean, one box per cat.
- The cat should have constant access to a warm nook for daily retreats.
- The cat should be groomed to maintain a healthy, clean coat.
- The cat should be kept free of parasites.
- Some long-haired and woolly cats can benefit from clipping in summertime.
- In any attempt to train a cat there should be no physical punishment.
- Certain limitations should be placed on any of the cat's activities that are not in accord with household standards. This can be done by prompt but humane discouragement applied consistently.

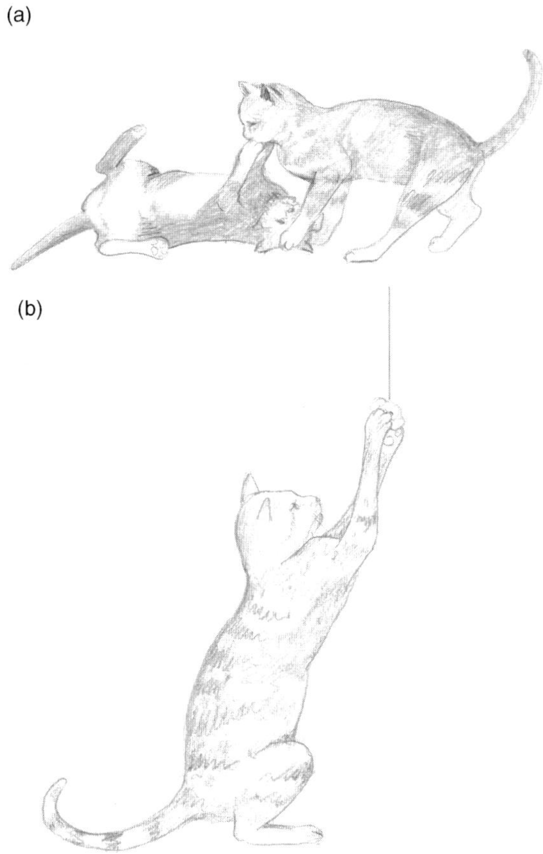

Fig. 13.1. (a) Partnered play; (b) solo play.

- A scratching post should be provided for the cat's innate clawing exercise.
- Frequent affectionate attentions should be paid daily to the cat by the caregiver and cat toys can help to occupy the animal's periodic interest in play, such as a table-tennis ball.
- Every cat routinely permitted outdoors should be fitted with a collar with an identification tag and possibly a small bell. They should not be put outdoors at night.
- Ideally, a small trap door should be fitted to an outside door for the exit and entry of a free-range cat in a rural area.
- Any cat lost for one day should be actively sought in the neighbourhood and at the local animal shelter.
- A pregnant cat near term should be provided with a secluded birthing box of adequate size.
- Future homes should be sought for unwanted kittens while they are still nursing.
- The destruction of any cat or kitten should only be done by euthanasia using an anaesthetic overdose or euthanizing drug.
- Vaccinations should be given in good time including any that are appropriate to the cat's locality.
- Veterinary attention should be provided promptly whenever any disorder, illness or injury might affect the cat. Evidence of pain requires immediate veterinary attention.

In addition to these welfare guidelines, there are several welfare warnings that are to be noted, as follows.

Welfare Warnings

- A housecat is bound to be in close proximity to any poisonous substances that may be kept in the household, shed or garage. This includes items that are in common use, such as aspirin, antifreeze, detergents, pesticides and paint materials. Even if a cat may not consume such items, its paws may become contaminated. On subsequent licking of its paws, the poison will be ingested.
- Housecats should not be put outdoors at night. It is cruel to do so in poor weather and it is dangerous for the cat to be wandering at night in any weather, since it is likely to encounter road traffic or dogs.
- Unvaccinated cats will be very prone to certain serious infections if allowed to encounter other cats.
- Radical and sudden changes of food are likely to cause digestive disorders that can be serious.
- Since cats may have a habit of following people in the household, doors should not be shut quickly behind a person moving through doorways, since cats sometimes get hurt severely by doors being closed on them.
- Unidentified cats that become lost are seldom recovered. Every housecat should have a collar and identification tag.

Unaddressed above are two very contentious matters relative to cat welfare. They concern excessive clawing and breeding, things that are the major detractive features of cat activity or ownership. They require to be closely examined as welfare problems since both are sources of much premature destruction of unwanted cats and kittens.

Excessive clawing

Cats have highly specialized claws that are normally withdrawn and enclosed in sheaths, but can be fully extruded in an instant. The claws are strong, needle pointed and curved hook-like. Such claws are ideal for catching prey, climbing, defending, fighting and making territorial marks, but have little purpose or need in house-bound living. If it is found necessary, the claws can be blunted by trimming. By pressing upward behind the toe pads, the cat's claws can be pushed out. This allows the tips of the claws to be clipped off. Great care must be taken to remove only the needle-pointed ends. The upper, thicker portion of the claw contains the 'quick', and any cut here will cause bleeding and considerable pain. Such an experience will never be forgotten by the cat and it will resist claw trimming from then on.

When kittens are born they already have their claws fully extruded and while they are suckling, they repeatedly press their little claws into their mother in a rhythmic, kneading fashion. Apparently, this is to stimulate the flow of her milk. Kittens become able to withdraw their claws by 3 or 4 days of age, but they still extrude them during suckling for this kneading activity. This kittenish, in-born, action pattern stays with them into mature life and kneading is often reproduced in an adult while lying comfortably on the lap of its caregiver and being petted. Adult claws are long and sharp, however, and are likely to cause considerable discomfort to the handler in the course of such kneading which can penetrate clothing. Claw trimming is indicated in this situation.

The forepaws are more functional than the hind ones. To extrude, hone and clean their important front claws routinely, cats have an 'innate action pattern' in a clawing exercise that they perform frequently. The action pattern involves rhythmically repeated and alternate forelimb stretching and pulling on the briefly anchored claws that have dug into an upright surface. This is done vigorously and gives muscular exercise to the entire clawing apparatus. This VICE (vigorous, innate, clawing exercise) is effectively the same action that is used outdoors by tomcats in creating their territorial marks on tree trunks or similar wooden upright structures. (Some cat critics regard this behaviour as a real vice.) It is commonly regarded as 'claw sharpening', and to some extent this may be part of the function of this compound behaviour (Fisher, 1993). This action stretches the muscles that flex the toes.

As VICE is an action pattern, identical in all cats, the neural 'wiring' for it is genetically established in the cat's brain from birth. Whenever it is enacted, it is an automatic performance and is as much a fixed part of the cat as its tail, its coat or its meow. Within the home, the cat will chose to perform its VICE on an upright corner or edge of some item of furniture that has a surface into which the claws can be hooked. Any attempt to make a cat give up this instinctive practice is futile. Training will not remove the instinct. The cat cannot help doing it. All that can be done, by way of discipline, is to direct the cat to a suitable rope-bound scratching post. The post can be manoeuvred tactically in sustained redirecting. Free-range, rural cats could perhaps be put outdoors as soon as a VICE commences.

Training cats at the best of times is difficult and training adult cats to use only a scratching post can be difficult for some of them, especially when the cat is left alone during the day. Destruction of furniture can occur from VICE and an owner's intolerance of this can lead to the unfortunate animal being rejected. A radical, life-or-death solution for house-bound cats is de-clawing of the forepaws. After de-clawing, the cat will still continue its VICE, but no hooking can be done, no damage is made and the exercise is still obtained. Furthermore, no child can be scratched.

Particularly in the philosophy of animal rights adherents, de-clawing is regarded as mutilation and is claimed to be painful. The further claim is that it is unethical to make the cat defenceless: if this were so, it would not contribute to well-being. But there is another point of view. Are tooth extractions, or medical plastic surgeries in humans considered mutilations? Are gall bladder removals or appendectomies as painful as the conditions necessitating the operations? De-clawing is done by veterinary surgeons using general anaesthesia to ensure that the procedure is entirely painless. It is true that the cat may experience digital pain for a few days following the operation, but this can be dealt with by the administration of painkillers. Afterwards the cat is a better domestic pet in any household with intolerance of scratching. Of course, such a cat has no place on the street. De-clawing may ensure a reprieve from the hazards of the street and longevity for the housecat; this might seem a stronger realistic case than the ethics of not altering the animal's natural anatomy. However, differences of opinion will persist and will encourage more thinking about cats. Even welfare theory is divided on this issue of cat de-clawing. What is clear regarding welfare is the fact that de-clawing is totally unsuitable for cats regularly allowed outdoors, since they would be unable to climb or defend themselves effectively.

Neutering

Neutering is another major issue in cat welfare. The term is used to cover the operations for both male and female sex organ removal, namely castration and ovariectomy, respectively. Again, some call them mutilations and claim that they are painful operations. The term 'mutilation' is somewhat provocative since neither operation causes a visible change in the animal's appearance. Both operations should be done before puberty and are done by veterinarians on the animal under general anaesthesia, ensuring that they are painless procedures.

In spaying, the anaesthetized female cat has a short incision made through the abdominal wall. The ovaries and uterine horns are brought out and removed. The abdominal incision is then closed with stitches. In castration, the anaesthetized male cat has a small incision made into each testicle, which is then separated from its cord. The cuts on the empty scrotum soon shrink and are left to heal without stitches since the cuts are too small for suturing.

The neutered cat cannot breed. The female now has no source of sex hormone, never comes into heat, never seeks a tomcat and never has litters of unwanted kittens. Her sterile status allows her to have the life of a domestic pet without all the stresses of repetitive, prolific reproduction. All this is a great assurance for her welfare throughout life. The castrated male cat no longer produces testosterone, has no sex drive, does not wander to seek out breeding females, does not constantly try to go free-range, does not produce the miasma of tomcat odour, does not develop the greasy condition of 'stud tail', but becomes a likely subject to be a good pet. Once again these sexless characteristics provide insurance for the animal's welfare in a tolerant, caring home.

Taking all the reproductive chemistry and business out of the cat's life by gonadectomy leaves it to concentrate on living for its self. This is a real blessing for the animal, to be free of the reproductive drive that can override all the routine motivations concerned with self-care. The neutered cat of either sex has its reproductive drive put permanently into neutral. A more benign manner of living is bestowed on the fortunate creature by such beneficial surgery. Any reference to this as 'mutilation' – a term loaded with negative connotation – is totally misleading. This anaesthetized operation thoroughly justifies the ends. In fact, it can be argued strongly that it is not humane to leave household cats at the mercy of their sex hormones throughout life.

The castrated male cat, free of its testosterone, is fortunate in many ways. Feline testosterone too often has toxic-like side-effects such as constant aggressiveness, hostility in general and belligerence with other tomcats. As a result of these attitudes, wounds and injuries may be sustained. In pursuit of their breeding intent, encounters with traffic on the street are almost inevitable and these are often fatal. The more cats that are neutered, the better cat welfare is in general. The minority that will undoubtedly escape the procedure can easily do all the breeding necessary to maintain a full cat population.

Health Protection

Perhaps because they possess the property of acute awareness concerning themselves in their immediate and present circumstances, cats appear to experience illness and injury with a high level of self-containment. Protecting the cat against its diseases and dangers is a top priority in cat care. A state of health and well-being should be established in a cat through the system of care enunciated in this text. This requires constant protection with a further system, one of preventive practices. It is impossible to anticipate and prevent every ailment or trauma that can afflict a cat in the course of its common domestic living, but the principal ones can be identified and some action can be taken against their occurrence. Immediately, a distinction has to be made between indoor and partly outdoor cat care on the matter of their health risks. A closer look at these two methods of cat husbandry is now necessary in relation to preventive practices and states of well-being that are inherent in each method.

Partly outdoor method of cat husbandry

The traditional method of taking care of a cat has been to retain the animal indoors for much of the day and to release it outdoors for a spell of freedom, often for a duration of many hours or for the entire night. This is done on a regular basis, often regardless of weather or the type of environment, whether city, suburb or country. The perceived advantages in this method include the following.

Outdoor advantages

- The cat obtains natural exercise.
- It may be able to engage in natural predatory behaviour on rodents.
- It may use its outdoor sessions for excreting.
- It is allowed to fulfil a curiosity urge and explore.
- It may have social contact with other cats.
- It may satisfy any wild element in the cat's nature, including grass eating, climbing, hiding and running.
- The simple fact of providing true freedom may be an act of kindness.

This system has certain disadvantages, including those listed below.

Outdoor disadvantages

- The cat, as a warm weather animal, can suffer exposure in inclement weather if it has no access to efficient shelter.
- The cat can encounter a hostile dog and receive injuries from it.
- The cat can be involved in a cat fight.
- In contact with several other cats, exposure to various infections and parasitic infestations is likely.
- The cat may become lost or join a local colony of feral cats and become permanently undomesticated, returning to ancient instincts.
- The animal is at risk of involvement in a traffic accident. In a city this is a high risk.
- Exposure to an act of cruelty is an additional risk.
- The cat may ingest toxic material.

It is apparent that these items are significant threats to the cat's well-being, health and life, although some health protection can be provided by vaccination. Certain types of neighbourhoods and locations, such as inner city and proximal to heavy traffic, pose very high risk to a wandering cat. A secluded environment, however, could be safer for such an animal. Daytime liberty is safer than night-time. All night-time release is unsafe.

An unneutered cat, especially a male, is likely to be at greater risk by wandering and being inattentive to traffic while motivated for breeding. Nevertheless, some cats adapt to their outdoor territories, using them suitably throughout a normal lifespan. Others are less successful. The successful outdoor cats that survive the risks of outdoor life are likely to be intelligent animals.

Indoor method of cat husbandry

While cats were seldom kept permanently indoors in former times, this method of cat care has now become common for urban individuals. This change in cat husbandry coincided with this animal becoming the most popular pet in Western society. By the end of the 20th century, cats were estimated to outnumber dogs in Europe and North America. In pet-owning homes, where no one was at home during the workday, it was found that cats tolerated the isolation better than dogs. In addition, in such households, the pet owner often had limited free time for providing a dog with its necessary walking exercise. Cats were found to be more accommodating pets than dogs in the new domestic culture of both parents working and general urbanization by the human population.

Cat owners soon learned about the statistics relating to lost cats, their 'road-kills' and life expectancies for roaming cats versus home cats. Many owners responded by simply keeping the cat in the house, never allowing it to get out. Routine husbandry procedures for closing doors and windows became the system for ensuring the animal's secure custody. The method has several advantages, as follows.

Indoor advantages

- The cat is effectively in quarantine, never exposed to any carrier of infection.
- The cat cannot become a victim of a traffic accident, dog attack or entrapment.
- The cat does not become exposed to severe weather or toxic waste.
- The cat does not breed if unneutered.
- The animal does not become a stray.

- The animal is not likely to suffer injury from risky, extra-mural activity.
- The cat is likely to enjoy longevity.

This method of cat husbandry has certain disadvantages, as listed below.

Indoor disadvantages

- The indoor cat does not get the benefit of free-living exercises such as climbing and may climb curtains instead.
- The cat may not be able to run for a significant distance despite sprinting being in its hunting nature.
- The cat must be provided with hunting and catching toys to compensate for deprived hunting need.
- Escapes can occur and there must be prompt efforts to find and retrieve the cat which is, of course, totally inexperienced in free living.
- Most escaped cats remain in hiding close to home – sometimes for days.
- The indoor cat that has little house room may need to be taken out periodically for a walk on a lead and harness during a quiet time in the neighbourhood, e.g. at night (the cat may even walk the owner).
- Vaccinations and identification are still needed for the eventuality of escape.
- The lone indoor cat may need a compatible companion cat for its well-being. This requires proper choosing, e.g. two neuters of the same sex and the second younger than the first.
- If the indoor cat is stressed with loneliness, lack of space and chronic confinement, it may on rare occasions develop the mental disorder of feline dysphoria, exhibiting symptoms of abnormal addictive habits such as chewing clothing.

Feline dysphoria, with its symptoms, is a lasting condition. It is the major disadvantage in the indoor method of cat care. Otherwise, this is probably the best method for city cats from the point of view of health protection and longevity, given quality of accommodation and activity. At its best, this method can ensure the city cat's well-being, with all the assets considered.

Displaced Behaviour

Various acts of behaviour in cats that are considered abnormal are in fact simply displacement activities that are produced in frustration. When certain feline functions become motivated but are impeded, alternative actions can occur and may develop into habits of inappropriate behaviour. Such habits have an addictive basis once established and are difficult to eliminate. Among indoor cats, common examples include the following:

- wool- or cloth-sucking and -chewing;
- urine spraying;
- random excreting;
- curtain climbing; and
- furniture tearing and scratching.

The behaviour of sucking and chewing soft clothing material is displaced kitten feeding. When it occurs, it is usually in cats that are often alone in limited space, without significant social involvement. A technical term for this condition is feline orosthenia, meaning that it is an obsessive–compulsive mouthing activity in the cat. The type most often affected is the young, ordinary adult housecat confined in limited space without diversionary activities. These circumstances indicate that the condition is initiated by displacement of action through frustrated motivation. Many cases recover in time, especially if environmental quality is afforded to the animal. Some show a temporary improvement when treated with a tranquilizer, but anxiolysis is not considered a long-term solution to the problem (Fogle, 2011).

Although the cat is not a keenly social subject, it has a need for involvement in activities or social situations that can provide affiliative satisfaction (Fig. 13.2). Most affiliative circumstances can create such opportunities for interaction. If output of normal forms of behaviour such as interaction, investigation, play or visual variety is less than adequate (in activity, or below a functional threshold of action), there is an accumulation of frustration leading to some comfort-seeking in displaced activity. When the cat is not asleep or resting, it needs to experience some participation in a dynamic environment for physiological balance between inactivity and some behaviour that is natural.

In natural circumstances all male cats, domestic or wild, spray urine on to objects in their claimed territory to warn any intruding cat that this is privately occupied territory – the possession of an adult male ready to defend its homeland. This instinctive behaviour is hormone driven and dependent on testosterone. It is a good method of

Fig. 13.2. Affectionate bondings.

reducing fights between males, since feline males choose to live alone in their selected locations. Spraying is therefore normal behaviour in the male cat if it is not neutered by castration. (The penis of the cat points backwards and the male, like the female, crouches to urinate. If the male takes a horizontal position it can direct urine backwards in the form of a spray.) Every domestic male cat should be neutered before puberty to relieve it of all the behavioural burdens of feline, masculine duties. Feline neutering is a welfare requirement.

In rare instances female cats and neutered males can occasionally perform spraying. If such cats present a problem in a household, treatment with oestrogen may remove the problem for a period. If spraying develops in a household with cats, one cat has the problem of maladjustment in relation to its accommodation or its relationship with another animal that it perceives as an adversary.

Control of Illness

Parasites and pathogenic microorganisms are the principal agents of disease and ailments in cats. Accidents of ingestion, however, also constitute a significant amount of health disorders in this animal. With their curiosity and investigative drive, cats tend to inspect novelties in their living range and then sample small items by mouthing them empirically.

Cats are particularly drawn to rubber bands and stringy strands; they will try to catch them with their forepaws and then bite down on a piece of the item. If this substance is firmly secured they pull on it; if it is free they may draw more into the mouth for the important test of 'mouth-feel'. Usually, any foreign substance is dropped from the mouth, but infrequently it is swallowed. The material is most likely to be swallowed if it is thread or thread-like. If it is long, ingestion will continue. If the thread is linked to a needle, the cat may try to swallow that also – and may manage to do so. String or thread becomes matted in the gut and an impaction can result. In every case of intestinal foreign body obstructions, prompt surgical intervention is needed.

The other occasional ingestive accident is the consumption of a toxic substance. A wide range of poisonous materials are known to affect cats by innocent but unfortunate intake when one of these has been made available to the animal by accident. Rodenticides, herbicides, arsenicals, antifreeze, phosphates, domestic cleaners and poisonous plants can be implicated in toxic incidents of cats. Urgent veterinary attention is necessary in such cases. Lavage of gastric content and intravenous fluid therapy can support the use of any antidotal treatment that might be indicated. In a number of cases the prognosis is uncertain, especially if the actual toxin, its quantity and the time of ingestion are uncertain factors. Some common indoor plants can be very toxic to cats, including spider plants, cyclamen, umbrella plants, poinsettia, foxglove, holly, caladium and daffodils.

Flowers can be toxic if chewed; these include monkshood, hyacinths, cornflowers and delphiniums. In addition, various items that are commonplace in the home are toxic to cats, including chocolate, fish oil, cocoa, powdered onion, aspirin and other medicines for humans. Sources of potentially hazardous substances should be realized and precautions taken against them (Table 13.1). If eaten, pieces of rodent cadaver containing poison such as strychnine, arsenic, phosphorus, warfarin, bromethaline, cholecalciferol and sodium fluoroacetate can cause secondary poisoning in cats. The use of gastric lavage with the administration of saline cathartics and the use of super-activated charcoal can be the basis for treatment, but the prognosis is often poor. Emetics may be contraindicated in some cases. Supportive therapy to maintain vital functions is the main rationale of treatments for poisoning in cats. Prevention is clearly the realistic manner of dealing with this threat to cat health. Examination of every object in its territory is behaviour that is lodged in the cat's nature. While this is desirable for the control of vermin, it can lead to misadventures.

Cats have changed little since their domestication. They still steal food, even when well fed. They still mark their homes by scratching. They climb and leap, and appreciate sunshine and warmth. They still practise hiding, hunting and capturing either real prey or toys. The kitten, however, is inevitably removed from the seclusion of the birth site and put into an entirely different environment. Although its own immune system has started to function, the juvenile kitten is faced with health challenges due to its susceptibility to serious infections resident in the broad population of cats. Even if the juvenile animal is being reared carefully as a house-bound cat, it will continue to remain exposed to feline viruses that may, or may not, respect closed doors. Outside these doors, the growing cat is exposed to all the hazards of infection, infestation and injury. Vaccination is vital to all cats in this age category and anti-parasite treatment should be scheduled routinely. A commonly recommended age for neutering is 4 months.

The practice of early socialization by sensitive handling (ESH) must be carried out thoroughly throughout the second month of life and into the third. This is one of the principal requirements in applied cat welfare. A phenomenal sensitivity exists

Table 13.1. Sources of hazards in domestic facilities presenting danger to cats.

Common location	Common hazards
House	Medicines, cosmetics, dyes, decayed food, abrasives, disinfectants, cleaners, bleach, detergents, moth balls, houseplants, insecticides, rodent poison
Garage	Oil fuel, kerosene, antifreeze, lead paint, paint stripper, wood preservative, glue, fillers, wood stain
Shed	Insecticides, herbicides, rodent poison, fungicide, slug bait, wasp nest, fertilizers, lime

in the kitten at that time, giving it lifetime acceptability of close human contact and proximity, if it experiences such contact then. The tactility inherent with sensitive handling seems to break down a strange feline factor of integumental hypersensitivity, such as is shown in the Jekyll and Hyde reaction. Without the experience of ESH, the cat is liable to have social problems in later domestic life relating to being handled. For example, it does not become a lap-cat. It is likely that those cats that avoid contact with humans or other animals, or even appearing aloof, may be ESH deprived. This period of social sensitivity ends after the third month has passed and, with its passing, occasions of tactility should now be increased by affectionate contacts on a fairly frequent basis to reinforce the socialization effect. Even if these contacts are brief, they serve a psychological purpose by compensating the socially sensitive kitten for its loss of contact with its mother and siblings, although such separation is natural to the animal in its physical and behavioural developments.

Juvenility is the main period for learning. If the kitten has not yet been introduced to a litter box, it must now be provided with one containing good quality, fine litter to a depth of at least 5 cm. Learning to use a litter box comes easily to a juvenile cat, as this behaviour is well implanted in its nature. A kitten that has not adopted a litter box immediately can be easily trained to use it, by repeatedly placing the animal on top of the clean litter whenever it is seen to start posturing for excreting. Good toileting behaviour helps a cat to be kept in favour in a household, so its welfare is more assured when it performs well at the litter box.

The use of a scratching post can be taught by placing the kitten at the post during the initial onset of scratching. Here again, disciplined scratching is appreciated by cat owners and it is the owner's responsibility to train the cat to perform its natural scratching in an acceptable way. This teaching should be repeated as often as is necessary to complete the training. An outlet for the cat's VICE is necessary for the animal's well-being. Kittens can be trained to desist from unwanted activity by promptly spraying them with water from a distance when they are performing the act that is to be prevented. A water pistol or large syringe can be used for the purpose of directing a fine jet of water. If the jet is sprayed from a distance, the cat is unlikely to associate the discipline with its trainer or caregiver.

By this simple water method, cats can be trained to stay away from houseplants. The water jet method can also be used to prevent the kitten from scratching furniture and curtains. The juvenile kitten must be trained in before such bad habits become established. The training should be started as soon as the juvenile cat arrives in its new and permanent home. Discipline can be carried out through vocal discouragement or water spraying. Lessons are not well learned, however, when the cat has a hyperactive episode, which is more common to occur in extended kittenhood. Any tendency of the kitten to bite its handler can be quickly discouraged by the handler blowing sharply on the young cat's face.

During the youthful phase of a cat's life the animal is still in danger with regard to health and habit. All the necessary vaccinations should have been completed by this time. The other great concern is over sex. The youthful female cat is too young for pregnancy, and the male youth should not be allowed to develop strong masculine habits and personality that would be unsuitable for its domestic role. Neutering for both sexes should therefore precede puberty at its average time.

When cats are permitted outdoors, adventurous activities are undertaken. These activities include such extravagant behaviour as climbing trees, racing across streets, playing with maximal vigour that often resembles manic activity, fighting seriously with associates, pouncing on moving articles, exploring dark recesses and taking a keen interest in the great outdoors. These activities can be dangerous and the cat that has no outdoor experience requires to be kept within limits.

Occasional, physical interferences of the mature cat are not to be overlooked in its care. Tufts of hair may have to be removed, dirty ears may call for cleaning, claws may require clipping, teeth might need to be de-scaled, vaccinations may need boosting and parasite control could become necessary. Clinical interventions are more likely needed by cats on day-release from home, but clinical attention is never completely removed as an emergency need for any cat in any age group. With the continuing quarantined status of the house-bound cat, there is great assurance that its health will not face challenges from an outside source. Inner problems, however, can become emergent. For some middle-aged cats, the wear and tear of natural life can start to affect organs and tissues, while entirely new pathologies can develop in the animal's ageing

Fig. 13.3. Window inhabitant.

body. Previously hidden disorders can now take advantage of any decline in the cat's state of health, becoming an established ailment. Notice must be taken of any sinister changes in breathing, chewing, eating, drinking, moving or excreting. Special notice of urination should be observed, and any sign of difficulty in passing urine should be regarded as a need for emergency veterinary treatment. When a suspected abnormality is detected and then confirmed as a disturbance to the cat's well-being, clinical attention is in prompt need.

Careful feeding is needed in all age groups. Overeating is to be avoided and a reduction of protein in the diet can ease the strain on the kidneys of ageing cats. Hairballs, anal glands, teeth, kidneys, allergies, parasites and various other items can become sources of health problems. For the outdoor-bound, middle-aged cat, the world continues to present a further source of hazards, including misadventure, trauma, infection and poison. Constant monitoring of a cat's activities and whereabouts is an important aspect of welfare.

During its old age, the cat can benefit from a degree of intensive animal care. This means monitoring what it consumes and passes on a daily basis. Maintaining a clean litter box, grooming, petting, providing clean food and water, as well as a comfortable bed and a warm retreat are necessary aspects of care. Such intensive care is essentially the usual care, but practised more intently and with further attention to detail. For example, if the animal's appetite is failing, tiny portions of favoured food can be given. Matted hair would demand professional trimming and grooming, while constipation would require a mild laxative and sores would need antiseptic treatment. Not least of all considerations in extra special care is the old cat's need of sunshine for its well-being. To facilitate this, a comfortable resting place should be allocated in a window with sunlight (Fig. 13.3).

As with any age, clinical attention should be provided whenever there is a suspicion about health. The veterinarian might alert the cat's owner of any sinister challenge to their pet's well-being that may have arisen. A course of action could then be determined for the animal, taking into account potential suffering that could be imminent. It is extremely difficult and perhaps even impossible for every cat owner to be fully prepared emotionally for the death of the pet. Human–animal bonds can be very strong when the animal has shared the home and provided life-long companionship in return for the fond care given by its household. At the time of pet loss there is much grief, which, regardless of the cause, has little ease from any rationalization. Grief is often expressed through caregivers' private remembrance of the intangible, vibrant qualities of their own, privately bonded animal.

14 Review of Well-being

Suffering

The main force against cat well-being is the factor of suffering. It can result from pain, illness or emotional stress. Suffering in the form of pain can be acute or chronic, i.e. sharp or lasting (Short and Van Poznak, 1991), and can vary in psychological severity from frustration or extreme malaise, being a constant component of illness (Rowan, 1988; Hotchner, 2007).

In any form, suffering is a burden on the psyche – the spring support of the self. In the text of this book abundant evidence of the cat's self has been given, indicating comparative vulnerability to the various forms of suffering that it might experience. Suffering can be an overwhelming, negative form in human life, and has the same capacity in cats (Gregory, 2004).

The fundamental objective in good cat care is the prevention of significant suffering, i.e. any degree of suffering that impairs or destroys well-being in the animal. Injury, hunger, exposure, illness, repression, abandonment and any painful factor are obvious agents of potential suffering that can affect cats. Cats with access to an outdoor, urban environment are particularly at risk of a mishap. Such events are largely preventable in circumstances of home restriction, in which the cat is kept indoors at all times. Exceptions can be made, however, such as accompanying the animal on a leash, carrying it in a cat box or placing it in a securely screened area of a garden. Such a system of preventive management must be strictly applied to ensure the protection of a state of well-being in the cat. Escape may occur, and at such a time there is a risk of loss. Prompt and thorough efforts at recovery are necessary for the animal's welfare and its protection from suffering, if it does become lost.

Population excess

Without spaying, a day-released female cat will produce over a hundred kittens in an average lifespan of 12 years. Based on animal control estimates, about 90 of these kittens will not find homes. Sooner or later they will be destroyed. In many cases, some degree of suffering will be experienced by these unwanted animals before final destruction. Their suffering will be in such forms as hunger, traumatic experience and fearfulness. All of this is preventable by the elective surgery of spaying the animal when it reaches the appropriate age (Spain *et al.*, 2004).

Having been taken under our domesticated conditions, cats are no longer subject to the population controls of wild nature. These controls would include a shorter lifespan from different causes such as infertility and smaller litters due to lower levels of nutrition. Human ownership of cats creates a range of ethical responsibilities including population control by breeding prevention.

Countless kittens, destined for destruction, result from the reckless reproductive spree of free-breeding felines. The cat's breeding is entirely hormone driven. To prevent mass suffering among the fall-out of population excess, the simple solution is to eliminate the breeding hormone produced by the male or female gonads. Excision of these unnecessary glands is humane. With its brain freed from the compulsion to breed, a cat is relieved of a drive that detracts from its own well-being. Neutering is a requirement in proper cat care – it indirectly prevents suffering in potential generations of surplus issue.

Misadventure

In the lives of cats the most common sources of suffering are injury and illness. Although injuries are usually acquired outdoors, accidents from burns, cuts and crushes can occur at home causing physical damage. Similarly, cats kept indoors can develop such clinical conditions as tumours, hairballs, intestinal disorders, dental conditions and urinary problems, which will lead to suffering.

Cats that are regularly put outdoors are subjected to a much greater risk of experiencing serious accidents and infections. Road accidents, if they are not immediately fatal, are very serious to cats, resulting in major fractures and internal injuries. Recovery from such circumstances, if at all possible, would be associated with much suffering. In such cases prompt relief with euthanasia or professional veterinary treatment and care would be required as a humane course of action (Rochlitz, 2005).

Disease control

As a rule, by thorough and updated vaccination the immunized outdoor cat is well protected against the major feline infectious diseases. Its excursions and meetings with other cats should not result in its infection. In very rare circumstances a vaccinated cat can have a poor immunological response and remain susceptible to the disease it was vaccinated against. Such a cat would be like all other non-immune cats and, in the course of roaming, would be likely to catch one of the serious diseases residing in the broad feline population. These diseases are heavily charged with clinical suffering and have mortality rates at significant levels. Panleukopenia, infectious peritonitis and feline immunodeficiency are three common viral diseases of cats that are fatal, as is rabies.

Some cats can hide their clinical sufferings quietly, while others express them through behaviour and audible callings. In most cases there are visible changes in appearance, habits, appetite, thirst, secretion or excretion. When the cat is suffering from a disorder of function, the caregiver will generally observe an aforementioned characteristic of abnormality (Fraser and Quine, 1989; Flecknell, 2001). In addition to pain or malaise, the sick cat is typically depressed in its inability to perform its normal repertoire of activities. These several neurogenic conditions place an oppressive burden on the animal's psyche, creating literal suffering. Our appreciation of an ongoing state of misery in a helpless domestic feline makes it clear that we are morally bound to address its suffering appropriately by veterinary attention (Balcombe, 2009).

When suffering cannot be alleviated, with its basic causation being beyond control or elimination, it becomes evident that its continuation is unnecessary. The essence of cruelty to animals is the imposition of unnecessary suffering on them. We can recognize necessary suffering, such as giving birth or brief manipulation to aid the animal in its capture or restraint. The continuation of suffering without ethical or biological justification requires prompt, painless termination in euthanasia.

Geriatrics

When the cat has lived its natural lifespan, various organs and systems of the body begin to fail. These deteriorations are irreversible as a rule, although therapeutic treatments can help or alleviate pathological conditions in certain cases for a limited time. It has already been acknowledged that well-being is relative to the status of the individual cat. In the case of any geriatric feline, well-being is inevitably impaired to some degree and eventually eliminated (Houpt and Beaver, 1981; Broom and Fraser, 2007). Even dementia, lately, has been recognized in veterinary medicine among certain aged cats. The owner is usually able to see the point at which a geriatric cat's continuing life is one of unnecessary suffering, with total loss of its well-being. Unfortunately, many owners may not be able to face this reality with resolve.

The psychological bond between owner and companion animal does not wither with age, but strengthens with time. No decision to sever this bond can be taken easily on objective evidence of final physical deterioration alone. Realization, however, must prevail that a cat's life has a limited well-being and that terminal suffering excluded from nature's methods must be ended by euthanasia when that limit has been reached.

Euthanasia

When the time comes for it, the death of a cat should occur either naturally or be induced painlessly. In the past, many cats were killed by cruel and inhumane means. In modern times, methods of ending the life of a cat properly are available at every veterinary clinic. Older methods deserve no description, since the euthanasia approach of today eclipses all that came before. Through the intravenous injection of a modern anaesthetic drug, unconsciousness is virtually instantaneous. The veterinarian then verifies that the animal's heart has stopped beating before confirming death. This procedure provides ideal euthanasia and is now a standard practice.

On the death of a cat, bereavement will be experienced by caregivers and others who may have shared an attachment to the animal. Frequently, the sense of bereavement is substantial in spite of comments from friends minimizing the loss as just an animal. Such remarks are of good intent to ease evident grief, but are declared in ignorance of the reality of the bond's emotional nature.

The intensity of grief relates to the associative value of the lost subject, to the individual experiencing the loss. Imagery can compound the actual loss, and no third party can estimate the sensory impact of a pet's death on the owner. One piece of reality to bear in mind at such a time, that may help to assuage regret over the loss, is that every form of life has a beginning and an ending. We are always thrilled by the beginnings and are saddened by the inevitable endings of sentient life.

If a cat owner's grief is acute, it would be advisable for the person to receive counselling from a suitable, capable party (Fogle, 2011). Many who are upset by the loss of a cherished cat seek help and relief in the prompt adoption of another feline subject in need. Cats and kittens everywhere are dependent on adoption and custody by people who appreciate their kind. The lost cat cannot be replaced as a personality or unique individual, but can have a successor – a close or distinct relative, perhaps.

An adoption, even in memory of a departed cat, is to be considered a most humane act. Many cat lives hang by threads, yet each one has the inherent capability to be the substance of a human–animal involvement. The human–animal bond, recognized, identified and evaluated recently, aids life for both animal and owner. This discovery may be the cat's new function or purpose. For the modern cat, its ultimate place may be as a human affiliate in the network of live beings in our world (Serpell, 1986; Jennings, 1997; Podberscek *et al.*, 2000).

Animal Rights in Domestication

With regard to the rights of animals, there requires to be a compromise between the perceived natural rights of the animal subject and the realism of domestication. For example, the right of freedom is removed in domestication. On the other hand, the right to a natural lifespan remains for a companion animal in sound health. Where there is conflict between an animal's natural right in domestication and a valid human interest, it is obvious in whose favour the conflict will be resolved. An ethical balance between these two factors is the better outcome, of course. Possibly, with regard to the companion animals, we are on the brink of achieving ethical standards in their uses. Even at the present time we can see that, within domestication, certain animal rights do exist for cats.

Feline rights

Any civilized and rational person, of normal mentality and intellect, knows that it is not humane to harm a cat deliberately and thereby to cause it to suffer unnecessarily. Society at large has the same opinion, the same intuitive knowledge, namely that it is not right to be cruel to animals, and in this case to cats. Cats have, therefore, by human intuition, been given the right to have freedom from harmful human deeds directed at them intentionally. Cats even have that right in the laws of many countries. Out of our intuitive sense of duty, we protect them as far as possible, from injury and disease by giving them care. So they have been given by us the right to good care in our custody. Again, we have given them the various rights that go with domestication in our current civilization. These rights relate to such matters as health care, nutrition and responsible ownership. To some extent, these rights are evidence of the advances that have occurred in civilized thought, opinion and education.

Animal rights that are not established in law must be well considered in order to be durable in various circumstances and to be essentially realistic. They also have to be trivalent, i.e. of value to the type of animal, to the human owner and to the community. First for the animal's well-being, second for the owner's interest, and third for the satisfaction of civilized society in general.

Given these conditions, ten humane rights can be itemized. Any cat, wild or domestic, in a home, laboratory, zoo or circus in human custody, should have these rights, as follows.

1. *The right to have freedom from human cruelty.* Unfortunately cats do not have such freedom from human deviants or animals such as dogs. Nature has not become involved with animal rights, but civilized nations have legal statutes against cruelty to animals, including cats of course, in order to protect them from cruel acts which some humans can do to them.

2. *The right to have good general care.* Such care relates to proper attention to the cat's needs from the cat's caregiver/owner. All the several items of good care have been detailed already in earlier chapters. Inadequate care could put the owner in a position for prosecution under animal cruelty laws (see item 1 above).

3. *The right to health care.* This right relates to the domestic care necessary to maintain normal health, to methods of preventive health care and also to clinical veterinary care in the event of injury or illness. The owner is placed under the responsibility to ensure such care at all times.

4. *The right to nutrition.* To meet this right, the cat's ownership should know about feline food needs in general. The caregiver should also cater to the feeding idiosyncrasies of the individual cat. Suitable food is the most vital need for the cat that is in the hands of the caregiver. Over-feeding should be avoided.

5. *The right to responsible ownership.* In this regard there are various and obvious dos and don'ts. In the main this right relates to protective custody and security in general. Turning the cat out for the night is irresponsible; it would be faced with too many hazards. Day-release in a suitable locality is more acceptable; for example, drivers are better able to see street cats in broad daylight. With this in mind, an old driving law requires revision. At present, any driver who kills or injures a dog on the road must report this to the police, but is not required to do so in the case of a cat. This is now unacceptable.

6. *The right to suitable quarters.* Farm cats may be able to find living quarters in some part of the farm's buildings, but all other cats must be based in, or confined to, a proper home. They must have comfortable accommodation within the home where they belong. It has been stated that comfort-seeking is a prominent part of the cat's nature. Comfort is therefore a need in this animal. It is an ethical requirement in the caregiver to meet this need which is apparently regarded by every cat as its birthright, its natural right, its God-given right. Cats basically need a comfortable, warm nook where they can retire and sleep at will (Fig. 14.1).

7. *The right to humane and affectionate handling.* Attention-seeking is another major drive in pet housecats and they have this right as the deal implicit with us in their domestication. Good handling with periodic, affectionate stroking deals with this right. This maintains the 'human–animal bond' that duly develops between the cat and its human associate and is beneficial to both parties.

8. *The right to life while well.* As a companion creature, a healthy cat should not be killed if its owner finds that the animal's ownership has simply become inconvenient. We appreciate ethically that cats, as sentient creatures, have a right to live as long as they have health and well-being. This right is met by dutiful human care that is comprehensive.

9. *The right to euthanasia when terminally afflicted.* Any cat suffering in a clinical sense from a condition that cannot be relieved expeditiously, or at all, has a right through human compassion to be relieved of its suffering by euthanasia. Many unwanted cats and kittens are destroyed in our society. If they have no home, no caring ownership,

Fig. 14.1. Asleep in a nook.

no full health, no domesticated status, it is a sad truth that they cannot be in a state of well-being. A humane end to the material suffering that goes with abandonment in such a domestic animal may be ethically warranted if all efforts at finding it a home have failed. Euthanasia should, of course, be painless. The optimal method is lethal injection with an anaesthetic overdose or with a euthanizing drug approved by the relevant veterinary association.

10. *The right to express behaviour that is natural and of fundamental importance to well-being.* These relate to self-protection, defence and food consumption (see Table 14.1).

All wild feline species have a further biological right, namely the right of the species to exist. The extinction of many wild cat species is already underway, however. The Convention of International Trade in Endangered Species (CITES) has listed twenty-one wild cat species threatened with extinction by the fur trade. These include the following: Andean mountain cat, Asiatic golden cat, Asiatic lion, black-footed cat, caracal, cheetah, clouded leopard, flat-headed cat, Geoffroy's cat, Iberian lynx, jaguar, jaguarundi, leopard cat, marbled cat, margay, ocelot, oncilla, puma, rusty-spotted cat, snow leopard and tiger.

These species are under some protection by CITES member governments that require a licence for anyone to trade in them for non-commercial purposes. Poaching and illegal sales undermine this system of protection in many countries.

In addition to CITES, the IUCN maintains a check on species known to be threatened with extinction, keeping lists of species that are under differing degrees of threat. The primary list is of those termed 'Critically Endangered'. That list contains the following big cats: Siberian tiger, Amur leopard, Anatolian leopard, Arabian leopard, Asiatic cheetah, Asiatic lion, Eastern cougar, Florida panther, North-African leopard, South-China tiger and Sumatran tiger.

At this time, the list does not yet include the newly discovered and somewhat imperiled new species called the Sunda clouded leopard, discovered on the islands of Sumatra and Borneo. On each of these islands, these cats are considered to be two different species, but linked somehow. They have a similar spotted appearance and live very secret lives. A conservation plan is being developed for them, since their existence is under threat from loss of habitat from logging.

It is clear that the wild cats of the world are in danger of complete elimination and various schemes are in hand for their conservation. Re-introductions to the wild with captive-born specimens have had mixed results. Such individual animals have not received maternal tuition on capturing prey or lessons on territorialism. The most promising approach is the type of scheme reported by Fiona Sunquist (1997), which essentially involves the active participation of local people receiving financial aid. Sunquist (2001) asserts that conservation efforts outside of parks and reserves will become increasingly important for the survival of wild cats. It has been pointed out that tourism may be the factor that can save the big cats by showing that they can exist with people's interests. The cats can live freely, in their natural areas that can also serve as locations for tourism (Thornton, 2011).

Table 14.1. Basic behavioural underpinnings for well-being in wild and domestic cats.

			Roles in well-being	
Basic behaviour	Homeostatic features	Maintenance goals	Short-range	Long-range
Fight, threaten or resist	Aggression, agonistic acts and play	Self-assertion Defence of territory	Suppression of competitor Retention of territory	Control over resources of prey
Flight or avoid	Submissive withdrawal, hiding	Avoidance of threatening situations	Immediate safety	Preserve self-interest
Feed and drink regularly	Search, capture and ingestion of food/prey Consumption of water	Self-maintenance in respect of visceral needs. Self-sufficiency in nutritional needs	Ingestion to repletion Suppression of hunger	Building up body reserves and storage of available energy
Rest and sleep	Physical inactivity	Physiological restoration	Energy conservation	Health care

Cat rites

All domestic cats have certain inborn items of behaviour that are performed in a very deliberate manner. In such cases, the form of each act is similar every time it occurs (Fox, 1974; Milani, 1987; Turner and Bateson, 1988; MacDonald, 1992; Schneck and Caravan, 1993; O'Farrell and Neville, 1994; Tabor 1997; Frazier and Eckroate, 2008).

Such specific actions are inherited rituals that can be regarded as ancient cat rites. They deserve to be recognized and respected as performances that are fixed in the common psyche of our cats. As such, these rites can be appreciated as behavioural pieces of cat heritage, behaviour that all cats have retained in their genes for over a hundred thousand generations. They have not been snuffed out by evolution (Edney, 1999; Robins, 2011). How much biological value there is in such gems of behavioural inheritance, such dynamic antiques. Seven cat rites can be recognized as follows.

1. *Birth actions*. At the birth of each of her kittens the queen cat efficiently severs the umbilical cord with her teeth, dry grooms each kitten and eats the afterbirth. The birth site is clean as a result, but the mother will try to move the litter to a new site after a few days.
2. *Burial of excretions*. As already described, cats are normally particular with their excretions, burying them by scratching soil or litter over them very thoroughly. The cleanliness in this is obvious.
3. *Territorial sign-posting*. Male, female and neutered cats that are allowed outdoors place territorial signs by spraying their urine on to prominent items in their environment. Males also place scratch marks in their territory and leave uncovered faeces as signs of their range when they venture away from home. Their social domain is clearly demonstrated to other cats, even when cat territories overlap.
4. *Vigorous innate clawing exercise*. This action pattern with the acronym VICE has been described earlier. It is a very typical cat action which keeps a cat's claws up to scratch.
5. *Kneading with claws*. This is another action pattern and has already been described. It serves as a signal of pleasurable association when the cat kneads on the lap of a human associate.
6. *Circular or extended sleeping position*. As previously described, when a cat is preparing to go into a proper sleep it lies on one side and either curls up into a very exact and tight circular shape or lies out in an extended posture. True sleep follows the adoption of either of these positions and this may last about half an hour; sleeps can be repeated. The tightness of the posture probably gives security. In the extended position the cat's body is stretched to the limit allowing the intervertebral joints maximal separation and minimal articulation. Spinal rest is the result of this positioning. Limbs are also outstretched and this assists the posture to be sustained. Some cats favour the circular shape in rest and others habitually employ extension (Fig. 14.2a and b).
7. *Grooming mode*. The most frequent grooming routine is wiping the sides of the head and face with the moist paw. In addition, there is reactive social grooming by licking the hair or skin of an accepted associate that may be cat or human. This is obviously a token gesture of harmony in a relationship.

Although cats have been domesticated only for less than seven millennia, from their rites it is evident that they have ancient ways which they retain as their own properties. In their partnership with humans, they now acquire the properties of fair rights, as concessions from informed society.

Features of Feline Well-being

Norms of well-being

In order to make a comprehensive determination of well-being in any cat, four features that characterize this status can be considered as essential criteria. These are the norms of the condition:

1. Absence of morbid behaviour.
2. Normal appetite and thirst.
3. Good body condition for age.
4. Association norms.

Ten behavioural indicators of feline well-being

In addition, various positive indicators of cat well-being can be observed in the subject's behaviour. Ten of these are recognized. Each serves as a pointer to the probable existence of the condition and a substantial number of them would be highly significant of the state of well-being in the subject. These indicators are enumerated below in their estimated order of considered importance.

1. Bursts of play.
2. Periodic pandiculation.

(a)

(b)

Fig. 14.2. (a) Tiger asleep in partly extended posture; (b) jaguar asleep in extension.

3. Grooming in all forms.
4. Positive croup-stroke reflex.
5. Face, temple and ear washing.
6. Attention-seeking and responding.
7. Wiping side of face or flank on caregiver.
8. Purring when petted.
9. Occasional, gentle vocalizations.
10. Resting on elevated sites.

Reaction

For the purpose of easily and quickly checking the condition of a housecat, the simple reaction of tail erection is quite useful. When the hand is stroked along the cat's back so as to bump against the root of the tail, the fit cat straightens its hindlegs slightly and raises the root of its tail firmly to oppose the stroke. This is a classic reflex. With frailty from age, poor condition or illness, this reflex is weak or absent.

Cats in good health should respond or react promptly and negatively to aversive events or changes in their environment. They should run from very loud noise or attack, hide from a pursuer, seek warmth when they experience cold and avoid sources of detectable harm such as stove tops. Similarly, healthy cats should respond positively to sources or items of reward, such as the offer of desired food, invitation to enter favourite places not usually available, and the provision of fresh litter.

The general image of the cat is of an animal possessing a lithesome quality of movement, being supple and agile in its body reactions. As a true hunter with keen self-interest, the cat is quickly alerted even when resting, should any significant stimulus occur to trigger an active response. It also produces lithe escape reactions when suddenly and firmly restrained. In its reaction times for most events, the cat has the greatest alacrity of all the domestic animals. The time lapse between stimulus and response in most circumstances is about a quarter of a second. This is almost immediate responsiveness.

Although they may slap an opponent with one forepaw, its claws fully extruded, cats are typically ambidextrous with their forepaws in fighting, in play or in catching prey. The feline ability to rotate the forearm contributes to the dexterity. Fast forepaw reaction is also linked to extreme accuracy in targeting the stimulus item with claw extruded. All of these are the characteristics of a predator on small and elusive prey.

Ingestion

If it has a stable existence and is in good care, the healthy cat should have a keen appetite. Preventive care should include the prompt removal of any item in the cat's environment that could be accidentally ingested. Because of a good, but curious appetite, a cat can sometimes ingest a harmful object. The cat's great trait of curiosity can spill over into various activities and result in such an undesirable swallowing accident. Good care has to take this possibility into account in the interest of the animal's well-being. Foreign body ingestion can occur easily and has serious consequences. In the interest of its well-being intervention should be prompt if the cat is seen to be mouthing any indigestible object.

Body care

What are the basic things that cats do to sustain themselves? They eat, hunt, bite, scratch, sleep, rest, explore, play, move athletically, use caution and seek comfort with security in human bonding. All through this range of occupational duties there is one other, major, feline characteristic – they keep themselves clean. The cat's various ways of keeping clean have already been described and they can be put under the term 'body care'. This prominent task in self-management is an outstanding protective feature of the healthy cat's behaviour. Its diligent activities in body care give it a hygienic screening against common body contaminants. Routine body care is a positive signal of well-being.

If it is unable to groom out any violation of its orderly body surface, any cat that is soiled or has a disordered coat does not have emotional well-being (Fig. 14.3). In such a case the caregiver has the responsibility of cleaning the animal by washing or clipping out the flaw. If hairballs accumulate in the coat of a long-haired cat, they tighten up progressively to the point where they pull on the underlying skin, causing continuous discomfort or pain. Thorough clipping relieves the condition with the restored well-being being evident in the cat's subsequent free-and-easy movements.

Motion

As a feature of motion, cat play is essentially simulated hunting or fighting and is therefore composed of the elements of these two wild activities once essential in its ancient existence. We accept racing, chasing, climbing, pouncing, exploring, catching, clawing and biting as playfulness in our cats, but these are obligatory actions that have fixed 'wiring' in the cat's nervous system. They are stuck in the cat's make-up defying evolution and domestication. Perhaps because they still serve another purpose; they provide the cat with a good programme of aerobic exercise, acting towards its physical well-being. Being aerobic, these activities stimulate heart and lung function, in addition to improving muscle and joint condition (Fig. 14.4).

Fig. 14.3. In need of grooming.

Fig. 14.4. Play fighting.

With its thyroid gland in prime state, the young, healthy cat is highly motivated to produce the playing exercises frequently. Since this play contributes to the young cat's well-being, it is important to facilitate play with the provision of cat toys. In addition, episodes of play can be induced or enhanced with the participation of the caregiver. As the animal goes through the steps of ageing the play sessions become less frequent and are less intense, but occasional playing continues to be exhibited as long as the cat enjoys well-being.

The cat's other manifestation of innately programmed exercise is pandiculation, which is anaerobic. Although this pattern of stretching is vigorous, even strenuous in cats, it does not involve a great output of energy and is involuntary. The heart and lung duo are not brought into this exercise, which is otherwise a singular contraction of the muscles that control skeletal movement overall or at either the anterior or posterior pole of the cat. It is a valuable, short exercise for the tone and particular flexibility of this little gymnast. Being involuntary, pandiculation cannot be induced or encouraged, but exercise with simpler forms of stretching in cats can be encouraged by teasing them with suitable lures. In the course of monitoring their condition, episodes of general or polar pandiculation are valuable indicators of well-being in cats.

Rest and sleep

Taking over 80% of the cat's lifetime, rest and sleep combined must rank with eating as the animal's top priority. Rest is usually taken in an upright position while sleep is taken lying flat on the side, either in an extended posture or an annular one. Sleep averages 18 h per day and is taken on more than one bed. These facts summarize the cat's method of body conservation and restoration. Sound health is based on this animal's uniquely thorough system of harbouring energy. This feline restfulness must be permitted, facilitated and noted in the course of ensuring well-being in the cat. Occasionally a cat will waken with a negative disposition that can last for hours.

Convalescence is a process of health restoration that receives too little notice in terms of animal care. Major requirements in convalescent care are rest and sleep. It may be that the cat is sometimes healing itself from minor dispositions in its protracted somnolent phases. It is common for cats to have minor dips in the usual well-being. Psychological disturbances can occur in cats, for example. Well-being is not necessarily a constant possession of even a physically normal cat. If it experiences stress at home, it may have temporary aversion to its domesticated niche, as already mentioned. Long phases of unconsciousness in sleep might allow reinforcement of a sense of security to return. This is speculation but the cat can have the benefit of the doubt. In all events, the cat's capacity for somnolence is an asset in health and can be a strong hint of its well-being, if the schedule is normal.

Association

Although we can see abundant dynamic evidence of a wild factor in our cat's occasional behaviour, its connection with our civilization is strong now. This association has already been described and termed the cat's 'pax domestica'. The relationship is certainly peaceful and resembles a truce between us and them, in which both parties benefit amiably (Karsh and Turner, 1988; Turner, 1988, 1991).

As a fairly sociable creature the typical housecat enjoys company (Ban, 1994). Naturally it prefers the company of its own kind, but it readily befriends alien cohabitants in the confines of its home. It particular, it adopts its human associates as its family. With this type or relationship there are the usual paradoxes, such as firm alliance and brief disputation, dependence and independence, physical connection and solitary withdrawal, and so on. This is a two-way linkage in which the caregivers/owners usually regard the cat as being in the family membership.

This familial relationship is the essence of the cat–human bond. This bond is now regarded as an occasional aid to human equanimity. Some of the physiological benefits that can be derived from this bond, while largely intangible, are believed to be real. The current, professional researchers into the phenomenon of the human–animal bond have produced data showing real benefits from it, such as reduced blood pressure, mood control and mental stimulation in the elderly. Emotional satisfaction from giving care to a dependent animal being can be great. The latter represents psychological gain and to this can be added the simple comfort obtained from the companionship of an amiable, enigmatic companion.

In their domestic associations much tolerance develops, as evidenced in the way that they accept other cats in the home. In some cases there is clear affection between pairs of neutered males (between play fights). Females also can have bonds with other cats that illustrate affection. Most housecats develop true affinity with their ownership. The relationship between entire tomcats and female cats is not in the tomcat's nature, as shown among the wild species. As previously asserted, all housecats should be neutered at an early age for their own well-being.

Exploration

In the desert any cat went without food if it did not explore its environment and investigate every recess and every slight movement or sound in it. If it was not curious by nature, a desert cat was not likely to discover any of the scarce items of nourishment in its habitat. Constant, keen curiosity about concealed things in the cat's world was a vital property for its survival. The jerboa and the cat had a tactical balance by one hiding and the other exploring. Evolution selected those cats that were the most assiduous investigators and were also versatile enough to seek prey on the ground, in the air and in water. This combination of exploration and versatility no doubt led them to an alternative manner of living by scavenging around the growing human populations in the villages by the eastern Mediterranean coast.

As the human population began to increase in the Middle East as a result of nutritional benefits created by the start of farming, the communities of the fertile crescent of the Near and Middle East expanded, with the inevitable increase in domestic refuse around them. This was like bait to the rats and cats of the adjacent desert lands. The tactical population balance in cats was gone and the increased numbers of cats duly found themselves in the human world. They still had the genes of vigilant, predatory behaviour for life. These were deep rooted by that time and are still there, including those for exploration and suspicious investigation. They are still useful traits by guarding the home against pest and invasion.

Territoriality

From the sign-posting activities that have been described in free-ranging cats, it is clear that each cat's home range is of great importance to it. All the work of scratch-marking, spot-spraying, scent-laying and patrolling that is performed by day-release cats shows preoccupation with real property and an appreciation of it. A desire by the cat to have its existence known by other cats can be assumed by its territorial rituals, as previously described. Two cats meeting together in the home range of one does not necessarily result in fighting; they can show mutual acceptance. On its home range, the domestic cat becomes itself, catching prey, such as birds, not really needed for food.

An important feature of the territorial behaviour of cats is their occasional tendency to roam beyond their home base. This walkabout habit can emerge in some cats unexpectedly and can cause them to travel beyond the home area with which they are familiar, almost in an exploratory manner. In their desert days, such wanderings could have led to the discovery of fresh sources of prey and would therefore improve their survival prospects. With their survival, individual cats possessing the innate tendency to roam ambitiously would likely have progeny with the same feature. This genetic basis of this behaviour would, in due course, become established in the broad population if cats in such an environment of limited and dispersed resources as the Libyan and Western deserts (Smithers, 1968; Clutton-Brock, 1987; Edney and Bessant, 2001).

Home-confined cats are obviously in a very different set of circumstances with regard to territorialism, but they have habits about their use of available space. They make use of windows that provide them with good visual fields, they have hiding places, other places for sleeping and favoured locations for perching or crouching. In addition, their travels about the home involve every part of the accommodation that they can use. Any places that they are forbidden to enter under normal circumstances become targets for them when any chance of access is created – and they may not be easily removed. In time they show much less interest in such places. The cat confined permanently to its home readily adjusts to its accommodation and, in time, makes no significant attempt to go outdoors unless an exploratory opportunity is offered to it in the form of an open window or door. This can have a fatal consequence. Prevention of this is one of the most important items in urban cat care.

As seen from the core of its habitat, the cat's visual field is included in its territory. For example, all of the ground area that it becomes accustomed to seeing from a window in its home is taken as its territorial possession, although it may never have any physical presence there. As a result, if the home cat is male and observes an alien cat in its secondary territory, it is perceived as a challenge to its status. Being unable to attack the intruder, it is frustrated stressfully and behavioural anomalies can result. These usually take the form of territorial marking within the home by urine spraying and/or faecal depositing. This is a return to innate sign-posting activity. In such an event, access to the outlook should be prevented until the behaviour of territorial marking has ceased. Attempts at stopping this behaviour by discipline alone are not rational; the animal is simply acting instinctively.

It is a general rule that, in adversity, the domestic cat's behaviour reverts to basically wild forms that are derived from its ancient genotype. What has been learned in the acclimatization to domesticated life can be temporarily lost in an instant of psychological disturbances.

Hiding

All cats adopt hidden positions periodically, often when they are in a mood of hyperactivity. Typically, the chosen location is at floor level and under a convenient item of furniture, often one with draped sides. While mostly hidden, the cat will peer out from the hiding place, watching the movements of some associate. The duration of hiding is usually no more than a few minutes. It is obviously an item of innate behaviour that finds expression in certain times of highly motivated activity, including play. As previously stated, feline play is practice or rehearsal for serious events such as fighting or hunting. In the course of the latter, hiding in wait for potential prey would be a useful tactic and this was probably the original value in this inborn habit. Its occurrence is an indicator of available energy and health. Hiding may also be the end point of flight.

With all these ten indicators of well-being taken into account, it is apparent that healthy cats engage in many activities while pursuing a full feline life. Even when it is permanently confined in a home, for its own security, a cat can produce all the forms of its true nature making use of play to do so in many instances. These activities can give evidence of a cat's well-being and deserve recognition in assessing the dynamic measure of health in the animal. Its degree of adaptation to its home is also revealed in its comprehensive behaviour. The term 'happiness' has not been used in this text until now, but it can be assumed that a well-adapted cat, showing the state of well-being in its appearance and conduct, is a happy cat (Fogle, 2011).

Physiological conservation

In addition to its parts, the cat has a whole that warrants attention. For a single biological type, the feline line has had a very long history of continuity, perhaps for 15 million years (Sunquist and Sunquist, 2002). Its mode of living is apparently based on a physiological scheme that is essentially protein-consuming and energy-conserving. This scheme has clearly been successful. Saving its energy is the feline specialty.

Walking is the most energy-efficient form of locomotion and the cat's typical walking rate, both in the wild and in domestication, is slow enough that within each stride there is a short phase when three legs are bearing weight while it is moving. In addition, the cat's manner of ingestion is quick deglutition that saves in the work of mastication. About 80% of the feline lifetime is spent in sleep and rest, both body states of inactivity. Obviously, the feline ethos displays physiological conservation.

Apart from the dramatic nature of performing as a carnivore, with the attendant ferocity, the cat practices bodily economy as the outstanding feature in its manner of living and moving. For example, when it walks, three limbs are always weight-bearing. Cat life is economical and this may be the reason for its biological success, together with the security from its spectacularly camouflaged colorations and the nature of its coat. It is ironic that the latter's attractiveness to the human eye has resulted in the decimation of the wild species from hunting. The feline forms of life can be regarded as a special model in nature, one that deserves full appreciation. This is a family of animals with members domesticated, wild or feral, now in need of appreciation and of human conservation.

All around the tropical countries with extensive forests there is increasing activity in logging. At times this is done by the timber industry at other times forest land is cleared by livestock farmers to create grazing land for cattle. For example, clearings are being created into the rainforest of Amazonia with pasturage then established in such clearances (Figs 14.5 and 14.6).

The effect of this is invasion of jaguar territory, placing cattle close to the cat's natural habitat. Dealing with the tropical heat of the day, the cattle tend to enter the forest along its edge for rest and shade (Fig. 14.7). This places these animals at the cat's disposal. Most cats will avoid such livestock, but sometimes cattle become prey. The responsible cat is then usually hunted down. This is the scenario around the world, with native cats finding their essential territories taken over in livestock farming. These are areas never previously farmed and they are cat habitats that are now permanently lost.

Fig. 14.5. Cleared edge of rainforest; note burned tree stumps.

Fig. 14.6. Grazing at grassed edge of rainforest.

Fig. 14.7. Cattle in shade at edge of rainforest.

Epilogue

More and more cats are valued as pets and often there are two or more kept together. More pedigreed breeds are emerging as the aristocracy of the feline family. At the same time, millions of cats live as ferals, avoiding domestication. These great increases in feline status contrast with the losses among the 38 species of wild cats that are being reduced in numbers around the world, to the point of extinction in many cases.

With the widespread use of all-terrain vehicles and the availability of automatic firearms, the hunting of wild cats for sport or fur now reaches a level at which the entire elimination of them becomes immediately feasible. The survival of these animals is an imperative need; they comb excess from the biomass and check the populations of proliferating pests that are hostile to human existence. Not least of all, they give spectacular biodiversity to the world.

All these cats, both domestic and wild, are in need of improved understanding concerning their customary behaviour, health and well-being. They are creatures with a common lifestyle that calls for admiration. With 10 million years of ancestral history, they have served a cause for nature and have been notable agents in evolution. Now they need the quality of care and consideration that society can give them with modern knowledge about their natures and disorders. As valued companions and free-living spirits, they deserve our fullest respect.

Glossary

Abnormal behaviour: Behaviour that deviates from a defined, comparable norm. The norm may be a behavioural form typical for a given genotype, age group, sex, breed, or type.

Aboulia: Loss of will-power as a mental disorder (see 'Learned helplessness').

Action pattern: A series of connected behavioural actions having consistent form and sequence.

Adaptation: The evolutionary process of becoming better adjusted to the environment through genetic change.

Adaptive behaviour: A behavioural change to suit the circumstances.

Affective: Emotional factors in behaviour.

Aggression: Any purposive action of a cat towards another with the actual or potential result of harming it, dominating it or depriving it of resources.

Aggressive behaviour: Actions with the tendency to initiate a vigorous conflict.

Agitation: Non-directed emotional activity, or extremely restless behaviour.

Agonistic behaviour: Any behaviour associated with conflict or fighting between two individuals.

Animal care: A system of husbandry for animals secured individually or in groups, in which places specify standards on environment, feeding, hygiene, health, protection and appropriate handling.

Anomalous behaviour: (i) Irregular behaviour. (ii) Behaviour that is a variant of a normal activity but displayed abnormally, e.g. excessive self-grooming, chewing cloth.

Anorexia: Abnormal lack of ingestive behaviour, e.g. in severe illness with a depressed state.

Anxiolytic: A drug relieving anxiety.

Appetitive behaviour: Behaviour manifested during the initial phase of an activity which indicates a desire to attain a certain goal, e.g. searching for food.

Applied ethology: The study of animal behaviour conducted primarily for practical application or for understanding.

Aversion stimulus: A noxious or painful stimulus.

Avoidance: Non-involvement with aggressive challenge by positioning.

Barren environment: An environment of insufficient complexity for a cat's sensory needs.

Behavioural disorder: Manifestation of behaviour that differs from that of a typical healthy cat. Behavioural disorders may be temporary or chronic.

Behavioural display: Any behaviour that has or may have communicative function, e.g. oestrus display.

Behavioural pattern: An organized sequence of behavioural actions having a specific design, e.g. feline mating.

Binocular vision: Vision in which images of the same object(s) are projected on the retinas of both eyes simultaneously, e.g. good forward vision.

Bullying: Persistent aggression by one cat towards another.

Cachexia: Clinical weakness of mind or body from disease.

Capacitation: The state of becoming capable for a specific function.

Caterwaul: The shrill howl of a cat.

Circadian: Referring to cyclic rhythm, corresponding closely to a 24 h interval.

Cognition: A process of perception, reasoning and development of expectations, e.g. anticipation of events.

Cognizant: Having specific knowledge.

Competition: (i) The direct struggle between individuals for a limited supply of resources or environmental necessities. (ii) The common striving for living requirements such as food, space or shelter by two or more individuals.

Competitive behaviour: Behaviour manifested to attain adequate or preferential use of a limited resource, e.g. food, space, sexual partner.

Conation: The desire to perform voluntary action; the performance of voluntary action.

Condition: The general health and fitness of a cat.

Conduction (of neural impulse): The transmission of an excitatory signal from a neurone to a target cell or cells.

Congener: Associate of the same species.

Congenital: Existing from birth (with or without inheritance).

Consummatory act: An act that constitutes the termination of a given behavioural pattern.

Contactual behaviour: Maintenance of bodily contact.

Cope: The ability to deal with demanding circumstances.

Core area: A location heavily used by the cat.

Critical period: The infantile and maternal phases when the subject is most sensitive to specific environmental features and experiences, e.g. the period in the second month of kittenhood when the animal can acquire a permanent acceptance of human association if it experiences the 'social effect'.

Defecation: Elimination of faeces from the body.

Defensive behaviour: Behaviour performed to prevent or neutralize a real or perceived aversive stimulus. According to the circumstances, such behaviour may encompass aggression, avoidance or signs of appeasement and subordination.

Depression: In a behavioural sense, a state of severe emotional dejection and atrophied behaviour occurring in various clinical disorders.

Deprivation: Removal of needed substances (feed deprivation, water deprivation), perceptual isolation from any other being.

Disorientation: Failure of an individual to acquire and keep a proper spatial orientation in its own environment.

Displacement activity: An act that is a substitute for another that is blocked or frustrated.

Distress: An emotional state of a subject resulting from excessive fear, loss of companion or object with which it has a strong psychological bond, physical discomfort, food and/or water deprivation, pain, etc. (see 'Suffering').

Diurnal: Pertaining to daylight hours or recurring daily, e.g. feeding times.

Dominance: An individual animal is said to be dominant over another when it has priority in feeding and sexual behaviour, and when it is superior in aggressiveness.

Drowsing: Being in a somnolent state characterized by reduced consciousness.

Dysphoria: A state of unease, mental discomfort, maladjustment.

Ecology: The study of the relationships between animals and their environment.

Eliminative behaviour: Behaviour involved in the expulsion of faeces or urine from the body.

Emotional behaviour: Behaviour that indicates a high level of excitation to cope with a disturbing situation.

Empathy: The ability to understand or assess feelings of other beings.

Empirical: Based on experience of trial and error.

Epigenesis: The development of the individual though early growing stages in a set order.

ESH (early social handling): The application of humane handling repeatedly to a kitten in its second month of age to induce a permanent state of social acceptance of people.

Ethics: Responsible standards of moral conduct.

Ethogram: (i) A record of behavioural activities. (ii) An inventory of behaviour patterns typical of an animal or a species.

Ethology: The study of the behaviour of animals.

Euthanasia: A painless killing without fear, anxiety or pain on the part of the subjected cat (by derivation, it means 'easy death').

Experience: Retention of knowledge or information from previous events in the life of a cat.

Exploration: An investigative activity with the potential to acquire new knowledge of its environment.

Extensor muscle: A muscle that extends or straightens out a part of the body, notably the limbs.

Extensor system: The variety of muscles that extend the forelimbs and hindlimbs of the cat.

Feeling: A state of awareness concerning the cat's well-being or lack of same.

Fight: An aggressive social interaction involving exchange of forceful and/or potentially harmful actions through some means of physical contact.

Flexion: Bending movement of a body part or limb that reduces the inner angle of a joint.

Frustration: A state of emotion that is produced when a cat is blocked in its attempts to achieve a goal.

Gene: A region of the long DNA molecule that duplicates a specific inherited characteristic in the offspring.

Genome: The complete set of genetic material in the subject.

Genotype: The genetic make-up of an individual.

Gestation: The period of intrauterine development (9 weeks in the domestic cat).

Goal: (i) A commodity or condition capable of reducing or eliminating motivation. (ii) An incentive.

Grooming: An act of integumentary care, e.g. biting, scratching, etc. Grooming is subdivided into: (i) self-grooming – a cat grooms itself; (ii) allogrooming – one cat grooms another; and (iii) mutual grooming – two cats grooming each other simultaneously.

Group structure: The internal arrangement of a group, such as age differentiation, sexual composition and social subdivision.

Habit: A persistent pattern of behaviour that has been acquired.

Habitat: The surroundings and conditions in which a cat lives.

Habituation: The permanent weakening of a response as a result of repeated stimulation.

Helminths: Parasitic worms.

Homeostasis: A state of psychophysiological balance within an animal by means of vital control of maintenance.

Home range: The locality whereby an individual conducts all its principal functions.

Homologous behaviour: Behaviour in different species that is similar in form, e.g. pandiculation.

Hormone: A chemical agent produced in specialized tissue or endocrine gland and carried in the bloodstream to influence responsiveness.

Hypothalamus: The very important part of the forebrain vitally involved with homeostasis and is the physical basis of the emotions. It is the output stage for the limbic system and the autonomic nervous system.

Hypothermia: Abnormally low body temperature (chilling).

Imprinting: Very rapid learning in a newborn to recognize its mother and bond with her.

Indoor housing: A housing system in which a cat is kept continuously inside.

Infection: The invasion of the body by disease-causing microorganisms such as bacteria and viruses.

Infestation: The invasion of the body by parasites affecting the integument or intestines.

Ingestive behaviour: (i) Actions by which an animal takes substances into the body by eating and swallowing. (ii) Behaviour concerned with the selection and consumption of food and drink.

Instinct: An inborn tendency to behave in a certain way.

Integument: All the covering of the body.

Intelligence: Ability of a cat to learn to cope with new situations and deal effectively with its environmental circumstances.

Investigative behaviour: Behaviour of a cat that involves inspection of an object or surroundings.

Kinesis: Bodily movement, notably locomotion using the limbs.

Kinetic: Pertaining to locomotion.

Learned behaviour: Any action performed as a result of, or influenced by, experience.

Learned helplessness: A condition of immobility resulting from chronic frustration (see 'Abulia').

Learning: The process that produces adaptive change in an individual's behaviour as the result of experience.

Maintenance behaviour: Any behaviour through which a cat sustains its own physiological equilibrium by use of resources. In a broader sense, this term refers to activities required for essential physical and psychological comfort and well-being.

Maladaptive behaviour: Any behaviour that directly or indirectly diminishes well-being, or leads to health problems or causes dysfunction of any kind.

Maladjustment: Failure to be comfortable with circumstances (see 'Dysphoria').

Malnutrition: A state of extended inadequate nutrition caused by deficient or unbalanced diet.

Melanistic: Variety of cat, coloured black from a recessive gene.

Mite: A microscopic member of the order Acarina. Most mites are non-parasitic with the notable exceptions of *Demodex folliculorum* and *Sarcoptes scabiei*. They can affect the skin with mange or scabies. Ticks are larger members of Acarina.

Morbidity: A state of sickness, or suffering a severe disease.

Moribund behaviour: Behaviour an animal performs when it is dying.

Motivation: The urge to perform a given behavioural function. Motivation arises when some neural controlling mechanism stimulates an appropriate activity for the well-being of the subject.

Motive state: The behavioural manifestations of a given state of motivation, e.g. threat postures indicating the physiological condition of fight or flight.

Motor activity: A function resulting from the excitation of the musculoskeletal system.

Mutation: An alteration of the DNA at a gene in the embryo that results in a change of inherited characteristic in the offspring, differing from that of the parents.

Neonatal: Relating to the newborn and the early days of its life.

Neuroethology: The study of the control of behaviour by the nervous system.

Neurogenesis: Growth and development of nervous tissue.

Neurogenic: Caused by or arising in nervous tissue.

Neurohormone: A hormone produced by nerve cells and secreted into the circulation.

Neurohumeral: Involving both the nervous system and the circulatory system.

Neurophysiology: The physiology of the nervous system.

Neuroplasticity: (i) The ability in the very young subject to develop fresh neural links through novel experiences. (ii) In mature subjects some degree of lost neurophysical function can be restored through stimulation of the brain by repeated attempts to perform the lost action, e.g. a three-legged cat acquires a new gait and a de-clawed cat acquires a new, rapid form of raking.

Normal behaviour: Behaviour that qualitatively and quantitatively does not deviate from regular or stabilized form. Commonly interpreted as behaviour of an animal that is healthy and free from a pathological condition.

Noxious: Unpleasant, painful, harmful or injurious.

Noxious stimulus: A stimulus that is unpleasant or harmful.

Odour: Sensation caused by chemical stimulation of receptors in the mucous membranes of the nasal cavities.

Oestrus: Temporary and recurring state of sexual receptivity in female cats; the adjective is oestrous.

Orthopaedic: Relating to the maintenance, or restoration, of normality in joints with their bones and muscles.

Pair bond: A continuing and physically close relationship between two animals.

Pairing: A voluntary affinitive relationship between two individuals. Temporary pair formations can be observed between sexual partners.

Pandiculation: Systematic outstretching as an action pattern.

Parturition: The process of giving birth.

Pheromone: A substance secreted by one individual and received as odour in a second individual of the same species, releasing a specific reaction of behaviour.

Phonation: Expression by sound.

Photoperiodicity: Regular, cyclic alternation between periods of light and darkness. Especially featured in seasonal changes of light periods.

Pica: Abnormal appetite for unusual and often inappropriate feed.

Play behaviour: A set of activities experienced as pleasurable in themselves by the performing organism. Play behaviour is frequently social in nature, and may imitate serious situations but without experiencing serious consequences.

Preconditioning: Preparation of an animal to cope with changes in environment (social and/or physical). This preparation may involve exposure to novel feeds, familiarization with different environments, handling, transporting, etc.

Predation: A form of interspecies relationship in which the attacker (predator) kills and eats the victim (prey).

Raking: Clawing downwards or horizontally in repeated scratching actions with the extruded claws of alternating feet being embedded in substances that can hold them briefly; the limbs, in turn, are stretched in a dragging action. It is a 'fixed action pattern'.

Reaction time: The interval between a stimulus and a response in behaviour.

Reflex: An innate and simple bodily/behavioural response involving the central nervous system and occurring immediately after the stimulus that evokes it.

REM (rapid eye movement): The deep stage of sleep in humans during which the eyeballs make rapid, flickering movements beneath closed eyelids. This stage in the cat is characterized by twitching muscular movements.

Responsiveness: The capacity of a subject to respond to given stimulation. It may be measured by some assessment of response latency.

Resting: A behavioural state characterized by cessation or reduction of movement and lowered expenditure of bodily energy, in order to avoid or recover from exhaustion. It is often accompanied by a lowered level of alertness.

Restraint: Any technique used to temporarily discourage or prevent unwanted movement. Restraint is used for examination, surgery, convalescence, breeding and safe handling.

Reward training: A type of operant conditioning in which a reward (positive reinforcer) is directly contingent on the desired performance of the subject. According to the training objectives, the performance resulting in reward may be either a produced response or a withheld natural response.

Saltation: The sudden act of leaping/running without evident purpose or cause.

Seasonal breeding: Breeding that occurs exclusively and regularly during a certain part of the year when certain conditions of the daily ratio of light to dark stimulate reproductive physiology.

Sensitization: The process of becoming more responsive to a given stimulus with practice or a number of trials.

Sentience: Capacity for sensing or feeling.

Sociability: The tendency to seek and maintain the company of peers. This term is sometimes used to refer to a cat's attachment to humans.

Social acceptance: A positive or at least neutral attitude toward other individuals or groups.

Social adaptation: Conforming fully with behavioural standards of a given social environment.

Social behaviour: (i) Activities directed towards and influenced by other members of a social unit. (ii) The reciprocal interactions of two or more animals and the resulting modifications of individual action systems. (iii) Any behaviour caused by or affecting another animal, usually one of the same species.

Social dominance: Ascendancy of an individual over another individual(s).

Social effect: The state of social acceptance in a kitten when it has experienced early social handling (ESH).

Social organization: (i) Any more or less stabilized system of intra- and/or interspecies social relationships. (ii) An aggregation of individuals into a fairly well-integrated and self-consistent group in which the unity is based upon the interdependence of the separate individuals.

Social role: A pattern of behaviour that an individual is reinforced to adopt as a member of a group.

Social status: The position attained by an individual in its social group, dependent on interaction between this individual and other members of the group.

Social subordinance: Acceptance of the ascendancy of another individual(s).

Social tolerance: The ability to accept the proximity of other subjects when using some common resource.

Socialization: A process of mutual familiarization between subjects which, if successful, leads to full social integration and relatively stable social arrangement.

Society: A group of individuals organized in some socially cohesive assembly.

Sociophilia: A cat's fondness for approaching people in general.

Sociophobia: A cat's extreme fear of strange people, e.g. visitors to the home.

Steroids: A group of biochemicals that include the sex hormones, cholesterol and the corticosteroids.

Stimulus: Any property of the body's environment that evokes a response, e.g. vocal commands.

Stress: (i) The psychophysiological consequences of challenging, tense or noxious situations. (ii) The endogenous result of aversive, exogenous factors.

Stressor: Any stress-inducing agent, e.g. physical injury, fear-provoking stimulus, etc.

Stress symptom: Any sign or behavioural display indicative of the effect of a stressor. The most common symptoms are increased excitability, reduced appetite, displacement activities and lethargy.

Stretching: A muscular activity, characterized by brief, forceful extension of limbs, and/or other parts of the body. Stretching is considered to be a comfort movement (see 'Pandiculation').

Suffering: The state of a cat enduring a noxious experience, injury, disease condition, infirmity or severe deprivation (see 'Distress').

Symptom: Subjective evidence concerning the abnormal health condition of a cat.

Syndrome: A set of signs that occur together and usually indicate reliably the occurrence of a specific disease or behavioural or physical disorder.

Tensor muscle: A muscle that tightens or stretches a body part.

Threat: Indication of intent to harm directed toward a specific adversary(s). Functionally, threat provides an opportunity for the threatened individual(s) to resolve the situation by escape or avoidance.

Threat signal: Any indication of threat. Such signals may include an aggressive posture.

Tonic immobility: A state of locomotor inertia shown particularly in an unwillingness to make responses that involve complex, coordinated bodily movements.

Training: The acquisition phase of a conditioning process. The term is also used for progressive conditioning. The cat's attention is essential in its training.

Trial-and-error learning: A type of learning whereby an animal progressively adopts actions that are most successful for a given purpose.

Ungulate: An animal with hooves.

Unnecessary suffering: Any suffering that is not essential for vital needs.

Visual acuity: Capacity for distinguishing visual detail.

Visual field: The complete area, including all objects in it, visible to the eye at any given time.

Vocalization: Production of sounds by the vibration of vocal cords in the larynx. The sounds may be modified by the structures of the pharynx.

Volition: Cognitive process that deals with decision making and voluntary pursuit of objectives (exercise of the will).

Welfare: The provision of care and services for the animal's well-being (in this text).

Well-being: A state of good physical and psychological condition exhibiting harmony between the cat and its surroundings. The most reliable indicators of well-being are good condition and health, combined with manifestations of normal behavioural repertoires.

Zygomatic ridge: The prominent edge of the cheek bone, which forms the outer edge of the eye socket.

Bibliography

Adamec, R.E., Stark-Adamec, C. and Livingstone, K.E. (1980a) The development of predatory aggression and defense in the cat (*Felis catus*). II. Development of aggression and defense in the first 164 days of life. *Behavior and Neural Biology* 30, 410–430.

Adamec, R.E., Stark-Adamec, C. and Livingstone, K.E. (1980b) The development of predatory aggression and defense in the cat (*Felis catus*). I. Effects of early experience on adult patterns of aggression and defense. *Behavior and Neural Biology* 30, 389–409.

Armstrong, J. (1975) Hand-rearing black-footed cats at the National Zoological Park, Washington. *International Zoo Yearbook* 15, 245–249.

Armstrong, J. (1977) The development and hand-rearing of black-footed cats. In: Eaton, R.L. (ed.) *The World's Cats*, Vol. 3(3). Winston Wildlife Safari, Winston, Oregon, pp. 71–80.

Bagley, D.V.M. (2005) *Fundamentals of Veterinary Clinical Neurology*. Blackwell, Ames, Iowa.

Balcombe, J. (2009) Animal pleasure and its moral significance. *Applied Animal Behaviour Science* 118, 208–216.

Ban, B.D. (1994) Cats need friends, too. Examining aggressive behavior in the cat. *Journal of the American Veterinary Medical Association* 205, 946–949.

Barrett, J. (2012) Call of the wild; meet the native cats of North America. In: *Cats USA 2012 Annual*. BowTie Inc., Irvine, California, pp. 68–75.

Barrett, J.A., Shaikh, M.B., Edinger, H. and Siegel, A. (1987) The effects of intrahypothalamic injections of norepinephrine upon affective defense behavior in the cat. *Brain Research* 426, 381–384.

Batley, T.N. (1993) *The African Leopard*. Columbia University Press, New York.

BBC Earth News (22 January 2011) Two forms of world's 'newest' cat, the Sunda leopard. Available at: http://news.bbc.co.uk/earth/hi/earth_news/newsid_9369000/9369238.stm (accessed 27 April 2012).

Beaver, B.V. (1994) *The Veterinarian's Encyclopedia of Animal Behavior*. Iowa State University Press, Ames, Iowa.

Beaver, B.V. (2004) Fractious cats and feline aggression. *Journal of Feline Medicine and Surgery* 6, 13–18.

Beltrán, J.F. and Delibes, M. (1993) Physical characteristics of Iberian lynxes (*Lynx pardinus*) from Doñana, southwestern Spain. *Journal of Mammalogy* 74, 852–862.

Beltrán, J.F. and Delibes, M. (1994) Environmental determinants of circadian activity of free-ranging Iberian lynxes. *Journal of Mammalogy* 75, 382–393.

Beltrán, J.F., Rice, J.E. and Honeycutt, R.L. (1996) Taxonomy of the Iberian lynx. *Nature* 379, 407–408.

Berrie, P.M. (1978) Home range of a young female Geoffroy's cat in Paraguay. *Carnivore* 1, 132–133.

Bessant, C. (2004a) *Catfacts*. Feline Advisory Bureau, Tisbury, UK.

Bessant, C. (2004b) *The Kitten Whisperer*. Barron's Educational Series, Hauppauge, New York.

Bezuijen, M.R. (2000) The occurrence of the flat-headed cat *Prionailurus planiceps* in south-east Sumatra. *Oryx* 34, 222–226.

Bhattacharyya, T. (1992) A brief note on some observations on the breeding biology of fishing cat (*Felis viverrina*). *Tigerpaper* 4(2), 20–21.

Bjärvall, A. (1992) Lynx and reindeer management in Sweden. In: *The Situation, Conservation Needs and Reintroduction of Lynx in Europe*. Environmental Encounters No. 11. Council of Europe Press, Strasbourg, France, pp. 40–42.

Blackshaw, J.K. (1992) Feline elimination problems. *Anthrozoös* 5, 52–56.

Blakemore, C. and Van Sluyters, R.C. (1975) Innate and environmental factors in the development of the kitten's visual cortex. *Journal of Physiology* 248, 663–716.

Blood, D.C. and Studdert, V.P. (2000) *Saunders Comprehensive Veterinary Dictionary*, 2nd edn. W.B. Saunders, London.

Boissy, A. (1995) Fear and fearfulness in animals. *The Quarterly Review of Biology* 70, 165–191.

Boland, L. (2011) The heart of Harry's problem. *Cat Care* 2, 50–51.

Bourke, A. and Rendall, J. (2009) *A Lion Called Christian*. Broadway Books, New York.

Bowland, J.M. (1990) Servals: wetland cats. *Endangered Wildlife* 1, 4–5.

Bradshaw, W.S. (1992) *The Behaviour of the Domestic Cat*. CAB International, Wallingford, UK.

Brand, C.J. and Keith, L.B. (1979) Lynx demography during a snowshoe hare decline in Alberta. *Journal of Wildlife Management* 43, 827–849.

Breitenmoser, U., Slough, B.G. and Breitenmoser-Würsten, C. (1993) Predators of cyclic prey – is the Canada lynx victim or profiteer of the snowshoe hare cycle? *Oikos* 66, 551–554.

Broad, K.D., Mimmack, M.L., Keverne, E.B. and Kendrick, K.M. (2002) Increased BDNF and trk-B mRNA expression in cortical and limbic regions following formation of a social recognition memory. *European Journal of Neuroscience* 16, 2166–2174.

Broad, S., Luxmoore, R. and Jenkins, M. (eds) (1988) *Significant Trade in Wildlife: A Review of Selected Species in CITIES Appendix II, Vol. 1: Mammals*. International Union for Conservation of Nature and Natural Resources Conservation Monitoring Centre, Cambridge, UK.

Broom, D.M. (2006) Behaviour and welfare in relation to pathology. *Applied Animal Behaviour Science* 97, 71–83.

Broom, D.M. and Fraser, A.F. (2007) *Domestic Animal Behaviour and Welfare*, 4th edn. CAB International, Wallingford, UK.

Brugiere, D. (2001) Range expansion of African golden cat in Guinea. *Cat News* 35, 14–15.

Canfield, J., Hansen, M.V., Quasha, J. and Diamond, W. (eds) (2011) *My Cat's Life*. Chicken Soup for the Soul Publishing, Cos Cob, Connecticut.

Carlson, D.G. and Griffin, J. (1995) *Cat Owner's Home Veterinary Handbook*. Howell Book House, New York.

Carlstead, K., Brown, J.L. and Seidensticker, J. (1993) Behavioural and adrenocortical responses to environmental changes in leopard cats (*Felis bengalensis*). *Zoo Biology* 12, 321–331.

Caro, T.M. (1994) *Cheetahs of the Serengeti Plains*. University of Chicago Press, Chicago, Illinois.

Cat News (1989) Mongolia exports manul skins. *Cat News* 11, 20.

Chapman, B.L. (1991) Feline aggression. Classification, diagnosis, and treatment. *Veterinary Clinics of North America. Small Animal Practice* 21, 315–327.

Chapman, B.L. and Voith, V.L. (1990) Cat aggression redirected to people: 14 cases (1981–1987). *Journal of the American Veterinary Medical Association* 196, 947–950.

Clutton-Brock, J. (1987) *A Natural History of Domesticated Mammals*. Cambridge University Press, Cambridge, UK.

Clutton-Brock, J. (1993) *Cats: Ancient and Modern*. Harvard University Press, Cambridge, Massachusetts.

Cooper, L. and Hart, B.L. (1992) Comparison of diazepam with progestin for effectiveness in suppression of urine spraying behavior in cats. *Journal of the American Veterinary Medical Association* 200, 797–801.

Crawshaw, P.G. and Quigley, H.B. (1989) Notes on ocelot movement and activity in the Pantanal region, Brazil. *Biotropica* 21, 377–379.

Crawshaw, P.G. Jr and Quigley, H.B. (1991) Jaguar spacing, activity, and habitat use in a seasonally flooded environment in Brazil. *Journal of Zoology* 223, 357–370.

Crowell-Davis, S.L., Barry, K. and Wolfe, R. (1997) Social behavior and aggressive problems of cats. *Veterinary Clinics of North America. Small Animal Practice* 27, 549–568.

Crowell-Davis, S.L., Curtis, T.M. and Knowles, R.J. (2004) Social organization in the cat: a modern understanding. *Journal of Feline Surgery* 6, 19–28.

Cuffe, D.F.C., Eachus, J.E., Jackson, O.F., Neville, P.F. and Remfry, J. (1983) Ear-tipping for identification of neutered feral cats. *Veterinary Record* 112, 129.

Cunningham, J.G. and Klein, B.G. (eds) (2007) Neurophysiology. In: *Veterinary Physiology*. Saunders Elsevier, Amsterdam, pp. 60–100.

Davidson, C. (2011) *The Cat Care Handbook.* Quercus Publishing, London.

Dawkins, M.S. (2004) Using behaviour to assess welfare. *Animal Welfare* 13, 53–57.

de Almeida, A.E. (1976) *Jaguar Hunting in the Mato Grosso.* Stanwill Press, London.

de Almeida, A.E. (1986) A survey and estimate of jaguar populations in some areas of Matto Grosso. In: *Wildlife Management in Neotropical Moist Forests: Conservation Status of the Jaguar. International symposium, Manaus, Brazil.* International Council for Game and Wildlife Conservation, Paris, pp. 80–89.

Dehasse, J. (1997) Feline urine spraying. *Applied Animal Behaviour Science* 52, 365–371.

Delibes, M., Blázquez, M.C., Rodríguez-Estrella, R. and Zapata, S.C. (1997) Seasonal food habits of bobcats (*Lynx rufus*) in subtropical Baja California Sur, Mexico. *Canadian Journal of Zoology* 74, 478–483.

Dennett, D.C. (1991) *Consciousness Explained.* Little, Brown & Co., Boston, Massachusetts.

Dorland's Illustrated Medical Dictionary (2003) 30th edn. Saunders, Philadelphia, Pennsylvania.

Dosa, D. (2010) *Making Rounds With Oscar: The Extraordinary Gift of an Ordinary Cat.* Hyperion, New York.

Dunstone, N., Durbin, L., Wyllie, I., Rose, S. and Acosta, G. (1998) Ecology of the kodkod in Laguna San Rafael National Park, Chile. *Cat News* 29, 18–20.

Dyce, K.M., Sack, W.O. and Wensing, C.J.G. (1996) *Textbook of Veterinary Anatomy*, 2nd edn. Saunders, Philadelphia, Pennsylvania.

Eagly, A.H. and Chaiken, S. (1993) *The Psychology of Attitudes.* Harcourt Brace Jovanovitch College Publishers, Fort Worth, Texas.

Eaton, R.E. (1974) *The Cheetah: Biology, Ecology and Behavior of an Endangered Species.* Van Nostrand Reinhold, New York.

Eaton, R.L. (1978) Why some felids copulate so much: a model for the evolution of copulation frequency. *Carnivore* 1, 42–51.

Edney, A.T.B. (1992) *The Complete Cat Care Manual.* Dorling Kindersley, London.

Edney, A.T.B. (1995) *101 Essential Tips on Caring for your Cat.* Dorling Kindersley, London.

Edney, A.T.B. (1999) *Cat: Wild Cats to Pampered Pets.* Taschen, London.

Edney, A. and Bessant, C. (2001) *The Complete Cat Handbook.* New Millennium Books, Vancouver, Canada.

Eisen, S. (2012) Choosing a healthy purebred cat. In: *Cats USA 2012 Annual.* BowTie Inc., Irvine, California, pp. 76–85.

Elliott, J.P., Cowan, I.M. and Holling, C.S. (1977) Prey capture by the African lion. *Canadian Journal of Zoology* 55, 1811–1828.

Emmons, L.H. (1987a) Comparative feeding ecology of felids in a Neotropical rainforest. *Behavior, Ecology and Sociobiology* 20, 271–283.

Emmons, L.H. (1987b) Jungle cruisers. *Animal Kingdom* January/February, 22–30.

Emmons, L.H. (1988) A field study of ocelots (*Felis pardalis*) in Peru. *Revue Ecology (Terre Vie)* 43, 133–157.

Emmons, L.H. (1991) Jaguars. In: Seidensticker, J. and Lumpkin, S. (eds) *Great Cats.* Rodale Press, Emmaus, Pennsylvania, pp. 62–63, 116–123.

Emmons, L.H., Sherman, P., Bolster, D., Goldizen, A. and Terborgh, J. (1989) Ocelot behavior in moonlight. In: Redford, K.H. and Eisenberg, J.F. (eds) *Advances in Neotropical Mammalogy.* Sandhill Crane Press, Gainesville, Florida, pp. 233–242.

Estes, R. (1992) *The Behavior Guide to African Mammals.* University of California Press, Berkley, California.

Feldman, E.C. and Nelson, R.W. (2004) *Canine and Feline Endocrinology and Reproduction.* Saunders, St Louis, Missouri.

Feline Advisory Bureau (2009) Fabcats. Available at: http://www.fabcats.org (accessed 10 June 2011).

Ferreras, P., Aldama, J.J., Beltrán, J.F. and Delibes, M. (1992) Rates and causes of mortality in a fragmented population of Iberian lynx *Felis pardina* Temminck, 1824. *Biological Conservation* 61, 197–202.

Ferreras, P., Beltrán, J.F., Aldama, J.J. and Delibes, M. (1997) Spatial organization and land tenure system of the endangered Iberian lynx (*Lynx pardinus*). *Journal of Zoology (London)* 243, 163–189.

Fisher, J. (ed.) (1993) *The Behaviour of Dogs and Cats.* Stanley Paul, London.

Flecknell, P. (2001) Recognition and assessment of pain in animals. In: Soulsby, Lord E.J.L. and Norton, D. (eds) Pain: Its Nature and Management in Man and Animals. *Royal Society of Medicine International Congress and Symposium Series* 246, 63–68.

Fogle, B.R. (1997) *The Encyclopaedia of the Cat.* Dorling Kindersley, London.

Fogle, B. (2011) *Complete Cat Care*. Mitchell Beazley, Octopus Publishing Group Ltd, London.

Forkman, B.A. (2002) Learning and cognition. In: Jensen, P. (ed.) *The Ethology of Domestic Animals*. CAB International, Wallingford, UK, pp. 51–64.

Fox, M. (1974) *Understanding Your Cat*. Bantam, New York.

Frank, D. and Dehasse, J. (2003) Differential diagnosis and management of human-directed aggression in cats. *Veterinary Clinics of North America. Small Animal Practice* 33, 269–286.

Fraser, A.F. and Quine, J.P. (1989) Veterinary examination of suffering as a behaviour linked condition. *Applied Animal Behaviour Science* 23, 353–364.

Fraser, C.M. (ed.) (1991) Urinary system. In: *The Merck Veterinary Manual*, 7th edn. Merck & Co., Inc., Rahway, New Jersey, pp. 869–894.

Frazier, A. and Eckroate, N. (1990) *It's a Cat's Life*. Berkley Books, New York.

Frazier, A. and Eckroate, N. (2008) *The Natural Cat*. Plume, Penguin Group, New York.

Gaona, P., Ferreras, P. and Delibes, M. (1998) Dynamics and viability of a metapopulation of the endangered Iberian lynx (*Lynx pardinus*). *Ecological Monographs* 68, 349–370.

Gardner, A.L. (1971) Notes on the little spotted cat, *Felis tigrina oncilla* Thomas, in Costa Rica. *Journal of Mammalogy* 52, 464–465.

Glusman, M. (1985) Brain mechanisms and aggressive behavior in the cat. *International Journal of Neurology* 19–20, 163–185.

Grassman, L.I. Jr and Tewes, M.E. (2000) Marbled cat in Northeastern Thailand. *Cat News* 33, 24.

Gregory, N.G. (2004) Sickness and disease. In: *Physiology and Behaviour of Animal Suffering*. Blackwell, Oxford, UK, pp. 183–192.

Griffin, D.R. (1992) *Animal Minds*. University of Chicago Press, Chicago, Illinois.

Guggisberg, C.A.W. (1975) *Wild Cats of the World*. Taplinger, New York.

Hackett, S.N. (2012) Pedigreed distinction. In: *Cats USA 2012 Annual*. BowTie Inc., Irvine, California, pp. 6–35.

Hall, S.L. (1998) Object play by adult animals. In: Bekoff, M. and Byers, J.A. (eds) *Animal Play: Evolutionary, Comparative and Ecological Perspectives*. Cambridge University Press, Cambridge, UK, pp. 455–460.

Hart, B.L. and Cooper, L. (1984) Factors relating to urine spraying and fighting in prepubertally gonadectomized cats. *Journal of the American Veterinary Association* 184, 1255–1258.

Hassanain, M., Bhatt, S. and Siegel, A. (2003) Differential modulation of feline defensive rage behavior in the medial hypothalamic 5-HT1A and 5-HT2 receptors. *Brain Research* 981, 201–209.

Hayes, C.L., Rubin, E.S., Jorgenson, M.C., Botta, R.A. and Boyce, W.M. (2000) Mountain lion predation on bighorn sheep in the Peninsular Ranges, California. *Journal of Wildlife Management* 64, 954–959.

Heidenberger, E. (1997) Housing conditions and behavioural problems of indoor cats as assessed by their owners. *Applied Animal Behaviour Science* 52, 345–364.

Heimer, L. (1978) Limbic mechanism. In: Livingstone, K.E. and Hornykiewicz, G. (eds) *The Continuing Evolution of the Limbic System Concept*. Plenum Press, New York, pp. 95–187.

Hogan, J.A. (2005) Motivation. In: Bolhuis, J.J. and Giraldeau, L.A. (eds) *The Behaviour of Animals*. Blackwell, Malden, Massachusetts, pp. 41–70.

Horwitz, D. (1997) Behavioral and environmental factors associated with elimination behavior problems in cats: a retrospective study. *Applied Animal Behaviour Science* 52, 129–137.

Hotchner, T. (2007) *The Cat Bible*. Gotham Books, New York.

Houpt, K.A. (ed.) (1997) Progress in Companion Animal Behavior. *Veterinary Clinics of North America. Small Veterinary Practice* 27(3). W.B. Saunders, Philadelphia, Pennsylvania.

Houpt, K.A., and Beaver, B. (1981) Behavioral problems of geriatric dogs and cats. *Veterinary Clinics of North America. Small Animal Practice* 11, 653–652.

Houpt, K.A. and Zicker, S. (2003) Dietary effects on canine and feline behaviour. *Veterinary Clinics of North America. Small Animal Practice* 33, 405–416, vii–viii.

Houpt, K.A., Honig, S.U. and Reisner, I.R. (1996) Breaking the human–companion animal bond. *Journal of the American Veterinary Medical Association* 20, 1653–1659.

Hubbal, A., Martin, J.E., Lang, B.D.E., Giorgio, R. and Knowles, C.H. (2009) On the organization of olfactory and vomeronasal cortices. *Progress in Neurobiology* 87, 21–30.

Hubbard, A.L., McOrist, S., Jones, T.W., Boid, R., Scott, R. and Easterbee, N. (1992) Is survival of European wildcats *Felis silvestris* in Britain threatened by interbreeding with domestic cats? *Biological Conservation* 61, 302–308.

Hunthausen, W. (2000) Evaluating a feline facial pheromone analogue to control urine spraying. *Veterinary Medicine (Lanexa, Kansas)* 95, 151–155.

ICONA (1990) Status and conservation of the pardel lynx (*Lynx pardina* Temminck, 1824) in the Iberian peninsula. In: *Convention on the Conservation of European Wildlife and Natural Habitats*. Council of Europe Press, Strasbourg, France, pp. 1–19.

Iggo, A. (1982) Cutaneous sensory mechanisms. In: Barlow, H.B. and Mollon, J.D. (eds) *The Senses*. Cambridge University Press, Cambridge, UK, pp. 369–408.

Insel, T.R. (1992) Oxytocin – a neuropeptide for affiliation: evidence from behavioural, receptor autoradiographic and comparative studies. *Psychoneuroendocrinology* 17, 3–35.

Iriarte, J.A., Johnson, W.E. and Franklin, W.L. (1991) Feeding ecology of the Patagonia puma in southernmost Chile. *Revista Chilena Historia National* 64, 145–156.

Izawa, M., Doi, T. and Ono, Y. (1982) Grouping patterns of feral cats (*Felis catus*) living on a small island in Japan. *Japanese Journal of Ecology* 32, 373–382.

Jackson, R.M. and Ahlborn, G.G. (1989) Snow leopards (*Panthera onca*) in Nepal – home range and movements. *National Geographical Research* 5, 161–175.

Jedrzejewski, W., Schmidt, K., Milkowski, L., Jedrzejewska, B. and Okarma, H. (1993) Foraging by lynx and its role in ungulate mortality: the local (Bialowieza Forest) and the Palaearctic viewpoints. *Acta Theriologica* 38, 385–403.

Jennings, L.B. (1997) Potential benefits of pet ownership in health promotion. *Journal of Holistic Nursing* 15, 358–372.

Jensen, P. (ed.) (2002) Behavioural genetics, evolution and domestication. In: *The Ethology of Domestic Animals*. CAB International, Wallingford, UK, pp. 13–30.

Johnson, K.A. and Franklin, W.L. (1984) Ecology and management of the Patagonia puma (*Felis concolor patagonica*) in southern Chile. In: Roberson, J. and Lindzey, F. (eds) *Proceedings of the Second Mountain Lion Workshop*. Utah Division of Wildlife Resources, Salt Lake City, Utah, pp. 141–146.

Johnson, W.E. and Franklin, W.L. (1991) Feeding and spatial ecology of the *Felis geoffroyi* in southern Patagonia. *Journal of Mammalogy* 72, 815–820.

Johnson, W.E., Culver, M., Iriarte, J.A., Eizirik, E., Seymour, K.L. and O'Brien, S.J. (1998) Tracking the evolution of the elusive Andean mountain cat (*Oreailurus jacobita*) from mitochondrial DNA. *Journal of Heredity* 89, 227–232.

Johnson-Bennett, P. (2011) *Think Like a Cat*. Penguin Books, New York.

Jones, S. (2001) *Darwin's Ghost. The Origin of Species Updated*. Random House of Canada Ltd, Toronto, Canada.

Kahn, C.C. and Line, S. (eds) (2010) *The Merck Veterinary Manual*, 10th edn. Merck & Co., Inc., Rahway, New Jersey.

Karanth, K.U. and Sunquist, M.E. (1992) Population structure, density and biomass of large herbivores in the tropical forests of Nagarahole, India. *Journal of Tropical Ecology* 8, 21–35.

Karsh, E.B. and Turner, D.C. (1988) The human–cat relationship. In: Turner, D.C. and Bateson, P. (eds) *The Domestic Cat: The Biology of its Behaviour*. Cambridge University Press, Cambridge, UK, pp. 159–177.

Kerby, G. and Macdonald, D.W. (1988) Cat society and the consequences of colony size. In: Turner, D.C. and Bateson, P. (eds) *The Domestic Cat: The Biology of its Behaviour*. Cambridge University Press, Cambridge, UK, pp. 67–81.

Kingdon, J. (1977) *East African Mammals*. Vol. 3A: *Carnivores*. University of Chicago Press, Chicago, Illinois.

Kitchener, A. (1991) *The Natural History of the Wild Cats*. Comstock Publishing, Ithaca, New York.

Klemm, W.R. (1984) Neurophysiology of consciousness and behavioural physiology. In: Swenson, M.J. (ed.) *Dukes' Physiology of Domestic Animals*, 10th edn. Cornell University Press, Ithaca, New York, pp. 671–706.

Knol, B.W. and Egberink-Alink, S.T. (1989) Treatment of problem behaviour in dogs and cats by castration and progestagen administration: a review. *Veterinary Quarterly* 11, 102–107.

Knowles, R.J., Curtis, T.M. and Crowell-Davis, S.L. (2004) Correlation of dominance as determined by agonistic interactions in feeding order in cats. *American Journal of Veterinary Research* 65, 1548–1556.

Koehler, G.M. and Hornocker, M.G. (1991) Seasonal resource use among mountain lions, bobcats, and coyotes. *Journal of Mammalogy* 72, 391–396.

Kolb, H.H. (1977) Wild cat. In: Corbet, G.B. and Southern, H.H. (eds) *The Handbook of British Mammals*. Blackwell Scientific, Oxford, UK, pp. 375–382.

Kortis, B. (2004) *The Neighborhood Cats' TNR Handbook: A Guide to Trap–Neuter–Return for the Feral Cat Caretaker*. Neighborhood Cats, Inc., New York.

Kortis, B. (2007) *Implementing a Community Trap–Neuter–Return Program*. The Humane Society of the United States, Washington, DC.

Krieger, M. (2012) The power of a simple click. In: *Kittens USA 2012 Annual*. BowTie Inc., Irvine, California, pp. 71–73.

Krotewicz, M. and Romaniuk, A. (1995) Social interactions in cats: regional brain monoamine distribution in dominant and submissive cats. *Acta Neurobiologiae Experimentalis* 55, 271–279.

Lachman, L. and Mickadeit, F. (2002) *Cats on the Counter: Therapy and Training for Your Cat*. St Martin's Press, New York.

Lanier, D.L. and Dewsbury, D.A. (1976) A quantitative study of copulatory behaviour of large Felidae. *Behavioral Processes* 1, 327–333.

Levine, M.S., Hull, C.D. and Buchwald, N.A. (1980) Development of motor activity in kittens. *Developmental Psychobiology* 13, 357–371.

Levine, M.S., Lloyd, R.L., Fisher, R.S., Hull, C.D. and Buchwald, N.A. (1987) Sensory, motor and cognitive alterations in aged cats. *Neurobiology of Aging* 8, 253–263.

Leyhausen, P. (1979) *Cat Behaviour, the Predatory and Social Behaviour of Domestic and Wild Cats*. Garland STPM Press, New York.

Leyhausen, P. (1988) The tame and the wild – another Just-So Story? In: Turner, D.C. and Bateson, P. (eds) *The Domestic Cat: The Biology of its Behaviour*. Cambridge University Press, Cambridge, UK, pp. 57–66.

Leyhausen, P. and Falkena, M. (1966) Breeding the Brazilian ocelot-cat *Leopardus tigrinus* in captivity. *International Zoo Yearbook* 6, 176–178.

Leyhausen, P. and Tonkin, B.A. (1966) Breeding the black-footed cat (*Felis nigripes*) in captivity. *International Zoo Yearbook* 6, 178–182.

Lim, B.L. (1999) The distribution, food habits and parasite patterns of the leopard cat (*Prionailurus bengalensis*) in Peninsular Malaysia. *Journal of Wildlife Parks* 17, 17–27.

Little, S. (ed.) (2012) *The Cat: Clinical Medicine and Management*. Elsevier, St Louis, Missouri.

López-González, C., González-Romero, A. and Laundré, J.W. (1998) Range extension of the bobcat (*Lynx rufus*) in Jalisco, Mexico. *Southwestern Naturalist* 43, 103–105.

Lorenz, K.Z. (1952) *King Solomon's Ring*. Penguin Books Ltd, London.

Lousada, A. (1956) Rearing serval cats. *Country Life* 120, 330–331.

Lovallo, M.J., Gilbert, J.H. and Gehring, T.M. (1993) Bobcat, *Felis rufus*, dens in an abandoned beaver, *Castor canadensis*, lodge. *Canadian Field Naturalist* 107, 108–109.

Lucherini, M., Sana, D. and Birochio, D. (1999) *The Andean Mountain Cat* (Oreailurus jacobita) *and the Other Wild Carnivores in the Proposed Anconquija National Park, Argentina*. Scientific Report No. 5. Società Zoologica La Torbiera, Milan, Italy.

Lucherini, M., Soler, L., Manfredi, C., Desbiez, A. and Marull, C. (2000) Geoffroy's cat in the Pampas grasslands. *Cat News* 33, 22–24.

Ludlow, M.E. and Sunquist, M.E. (1987) Ecology and behavior of ocelots in Venezuela. *National Geographic Research* 3, 447–461.

MacDonald, D. (1992) *The Velvet Claw*. BBC Books, London.

Macdonald, D.W., Apps, P.J., Carr, G.M. and Kerby, G. (1987) Social dynamics, nursing coalitions and infanticide among farm cats, *Felis catus*. *Advances in Ethology (Supplement to Ethology)* 28, 1–64.

McMahan, L.R. (1986) The international cat trade. In: Miller, S.D. and Everett, D.D. (eds) *Cats of the World: Biology, Conservation and Management*. National Wildlife Federation, Washington, DC, pp. 461–488.

McVittie, R. (1979) Changes in the social behavior of South West African cheetah. *Madoqua* 11, 171–184.

Maeda, H. and Maki, S. (1987) Dopamine agonists produce functional recovery from septal lesions which affect hypothalamic defensive attack in cats. *Brain Research* 407, 381–385.

Manteca, X. (1995) Animal behavior case of the month. Aggression of 2 weeks' duration between 2 cats in the same house. *Journal of the American Veterinary Medical Association* 206, 317–318.

Marder, A.R. (1991) Animal bites: behavior modification of the offending animal. *Seminars in Veterinary Medicine and Surgery (Small Animals)* 6, 192–198.

Marder, A.R. and Marder, L.R. (1985) Human–companion animal relationships and animal behavior problems. *Veterinary Clinics of North America. Small Animal Practice* 15, 411–421.

Martin, P. and Bateson, P. (1985) The ontogeny of locomotor play behaviour in the domestic cat. *Animal Behaviour* 33, 502–510.

Martin, P. and Bateson, P. (1988) Behavioural development in the cat. In: Turner, D.C. and Bateson, P. (eds) *The Domestic Cat: The Biology of its Behaviour*. Cambridge University Press, Cambridge, UK, pp. 9–22.

Matlacak, C.R. and Evans, A.J. (1992) Diet and condition of bobcats, *Lynx rufus*, in Nova Scotia during autumn and winter. *Canadian Journal of Zoology* 70, 1114–1119.

Mendl, M. (1988) The effects of litter-sized variation on the development of play behaviour in the domestic cat: litters of one and two. *Animal Behaviour* 36, 20–34.

Mendl, M. and Harcourt, R. (1988) Individuality in the domestic cat. In: Turner, D.C. and Bateson, P. (eds) *The Domestic Cat: The Biology of its Behaviour*. Cambridge University Press, Cambridge, UK, pp. 41–54.

Mendoza, D.L. and Ramirez, J.M. (1987) Play in kittens (*Felis domesticus*) and its association with cohesion and aggression. *Bulletin of the Psychonomic Society* 25, 27–30.

Milani, M.M. (1987) *The Body Language and Emotion of Cats*. William Morrow, New York.

Mills, D.S. and Mills, C.B. (2001) Evaluation of a novel method for delivering a synthetic analogue of feline facial pheromone to control urine spraying by cats. *Veterinary Record* 149, 197–199.

Mills, D.S. and White, J.C. (2000) Long-term follow up of the effect of a pheromone therapy on feline spraying behaviour. *Veterinary Record* 147, 746–747.

Mills, M.G.L. (1990) The lion (*Panthera leo*) and cheetah (*Acinonyx jubatus*) in Kruger National Park, South Africa. *Felid* 4(1), 13.

Mills, M.G.L. and Biggs, H.C. (1993) Prey apportionment and related ecological relationships between large carnivores in Kruger National Park. *Symposium of the Zoological Society of London* 65, 253–268.

Miquelle, D., Merrill, T.W., Dunishenko, Y., Smirnov, E.N., Quigley, H.B., Pikunov, D.G. and Hornocker, M. (1999) A habitat protection plan for the Amur tiger: developing political and ecological criteria for a viable land-use plan. In: Seidensticker, J., Christie, S. and Jackson, P. (eds) *Riding the Tiger: Tiger Conservation in Human-Dominated Landscapes*. Cambridge University Press, Cambridge, UK, pp. 273–295.

Mondolfi, E. and Hoogesteijn, R. (1986) Notes on the biology and status of the jaguar in Venezuela. In: Miller, S.D. and Everett, D.D. (eds) *Cats of the World: Biology, Conservation and Management*. National Wildlife Federation, Washington, DC, pp. 85–123.

Morris, D. (1987) *Catwatching*. Crown Publishers, New York.

Murray, D.L., Boutin, S. and O'Donoghue, M. (1994) Winter habitat selection by lynx and coyotes in relation to snowshoe hare abundance. *Canadian Journal of Zoology* 72, 1444–1451.

Myers, L.J. and Coulter, D.B. (2004) Vision. In: Reece, W.O. (ed.) *Dukes' Physiology of Domestic Animals*, 12th edn. Cornell University Press, Ithaca, New York, pp. 843–851.

Nelson, R.J. (2000) *An Introduction to Behaviour and Endocrinology*, 2nd edn. Sinauer Associates, Sunderland, Massachusetts.

Nicol, C. (2006) How animals learn from each other. *Applied Animal Behaviour Science* 100, 58–63.

Norton, T.T. (1974) Receptive-field properties of superior colliculus cells and development of visual behaviour in kittens. *Journal of Neurophysiology* 37, 674–690.

Nowell, K. and Jackson, P. (1996) *Wild Cats: A Status Survey and Conservation Action Plan*. International Union for Conservation of Nature and Natural Resources, Gland, Switzerland.

Nuñez, R., Miller, B. and Lindzey, F. (2000) Food habits of jaguars and pumas in Jalisco, Mexico. *Journal of Zoology (London)* 252, 373–379.

O'Brien, S.J., Collier, G.E., Benveniste, R.E., Nash, W.G., Newman, A.K., Simonson, J.M., Eichelberger, M.A., Seal, U.S., Janssen, D., Bush, M. and Wildt, D.E. (1987) Setting the molecular clock in Felidae: the great cats, *Panthera*. In: Tilson, R.L. and Seal, U.S. (eds) *Tigers of the World: The Biology, Biopolitics, Management, and Conservation of an Endangered Species*. Noyes Publications, Park Ridge, New Jersey, pp. 10–27.

O'Farrell, V. and Neville, P. (1994) *The Manual of Feline Behaviour*. Blackwell Science, Oxford, UK.

Okarma, H., Jedrzejewski, W., Schmidt, K., Kowalczyk, R. and Jedrzejewska, B. (1997) Predation of Eurasian lynx on roe deer and red deer in Bialowieza Primeval Forest, Poland. *Acta Theriologica* 42, 203–224.

Oliver, J.E. and Lorenz, M.D. (1993) *Handbook of Veterinary Neurology*, 2nd edn. W.B. Saunders, Philadelphia, Pennsylvania.

Overall, K.L., Rodan, I., Beaver, B.V., Carney, H., Crowell-Davis, S., Hird, N., Kudrak, S., Wexler-Mitchel, E.; Panel on Feline Behavior Guidelines, American Association of Feline Practitioners (2005) Feline behavior guidelines from the American Association of Feline Practitioners. *Journal of the American Veterinary Medical Association* 227, 70–84.

Packer, C. and Pusey, A.E. (1982) Cooperation and competition within coalitions of male lions (*Panthera leo*). *Animal Behaviour* 31, 334–340.

Palomares, F., Rodríguez, A., Laffitte, R. and Delibes, M. (1991) The status and distribution of the Iberian lynx *Felis pardina* (Temminck) in Coto Doñana area, southwestern Spain. *Biology of Conservation* 57, 159–169.

Passanisi, W.C., Macdonald, D.W. and Kerby, G. (1991) Wild cats and feral cats. In: Seidensticker, J. and Lumpkin, S. (eds) *Great Cats*. Rodale Press, Emmaus, Pennsylvania, pp. 162–169.

Pellis, S.M. and Pellis, V.C. (1998) The structure function interface in the analysis of play-fighting. In: Bekoff, M. and Byers, J.A. (eds) *Animal Play: Evolutionary, Comparative and Ecological Perspectives*. Cambridge University Press, Cambridge, UK, pp. 115–140.

Perry, R. (1965) *The World of the Tiger*. Athenaeum, New York.

Peterson, P.K., Chao, C.C., Molitor, T., Murtaug M., Strogar, F. and Sharp, B.M. (1991) Stress and pathogenesis of infectious disease. *Review of Infectious Diseases* 13, 710–720.

Pitcairn, R.H. and Pitcairn, S.H. (2005) *Dr. Pitcairn's Complete Guide to Natural Health for Dogs and Cats*. Rodale Press, Emmaus, Pennsylvania.

Pocock, R.I. (1941) The races of the ocelot and the margay. *Field Museum of Natural History, Zoology Series* 27, 319–369.

Podberscek, A.L., Paul, E.S. and Serpell, J.A. (2000) *Companion Animals and Us: Exploring the Relationships between People and Pets*. Cambridge University Press, Cambridge, UK.

Poole, K.G. (1994) Characteristics of an unharvested lynx population during a snowshoe hare decline. *Journal of Wildlife Management* 58, 608–618.

Poole, K.G. (1995) Spatial organization of a lynx population. *Canadian Journal of Zoology* 73, 632–641.

Poole, K.G. (1997) Dispersal patterns of lynx in the Northwest Territories. *Journal of Wildlife Management* 61, 497–505.

Poole, T. (1996) Natural behaviour is simply a question of survival. *Animal Welfare* 5, 218.

Prince, J.H. (1977) The eye and vision. In: Swenson, M.J. (ed.) *Dukes' Physiology of Domestic Animals*, 9th edn. Cornell University Press, Ithaca, New York, pp. 696–712.

Prosser, C.L. (ed.) (1991) Environmental and metabolic animal physiology: neural and integrative animal psychology. In: *Comparative Animal Physiology*, 4th edn. John Wiley, Chichester, UK, pp. 353–436.

Pryor, P.A., Hart, B.L., Bain, M.J. and Cliff, K.D. (2001) Causes of urine marking in cats and effects of environmental management on frequency of marking. *Journal of the American Veterinary Medical Association* 219, 1709–1713.

Pryse-Phillips, W. (2009) *Companion to Clinical Neurology*, 3rd edn. Oxford University Press, Oxford, UK.

Quigley, H.B. and Crawshaw, P.G. (1992) A conservation plan for the jaguar (*Panthera onca*) in the Pantanal region of Brazil. *Biology of Conservation* 61, 149–157.

Quinn, P.J., Markey, B.K., Carter, M.E., Donnelley, W.J. and Leonard, F.C. (2000) *Veterinary Microbiology and Microbial Diseases*. Blackwell, Oxford, UK.

Rabinowitz, A. (1986a) *Jaguar: Struggle and Triumph in the Jungles of Belize*. Arbor House, New York.

Rabinowitz, A.R. (1986b) Jaguar predation on domestic livestock in Belize. *Wildlife Society Bulletin* 14, 170–174.

Rabinowitz, A. (1990) Notes on the behavior and movement of leopard cats, *Felis bengalensis*, in a dry tropical forest mosaic in Thailand. *Biotropica* 22, 397–403.

Rabinowitz, A.R. and Nottingham, B.G. (1986) Ecology and behaviour of the jaguar (*Panthera orca*) in Belize, Central America. *Journal of Zoology* 210, 149–159.

Reisner, I.R., Houpt, K.A., Erb, H.N. and Quimby, F.W. (1994) Friendliness to humans and defensive aggression in cats: influence of handling and paternity. *Physiology of Behavior* 55, 1119–1124.

Robertshaw, D. (2004) Temperature regulation and the thermal environment. In: Reece, W.O. (ed.) *Duke's Physiology of Domestic Animals*, 12 edn. Cornell University Press, Ithaca, New York, pp. 962–973.

Robins, S. (2011) Wild cat to house cat. In: *Cats USA 2011 Annual*. BowTie Inc., Irvine, California, pp. 48–53.

Robinson, N.E. (2007) Homeostasis. In: Cunningham, J.G. and Klein, B.G. (eds) *Veterinary Physiology*. Saunders Elsevier, Amsterdam, pp. 620–639.

Rochlitz, I. (ed.) (2005) *The Welfare of Cats*. Springer, Berlin.

Rodríguez, A. and Delibes, M. (1992) Current range and status of the Iberian lynx *Felis pardina* Temminck, 1824 in Spain. *Biology of Conservation* 61, 189–196.

Rollin, B.E. (ed.) (1995) *The Experimental Animal in Biomedical Research*, Vol. II. CRC Press, Boca Raton, Florida.

Rowan, A.N (1988) Animal anxiety and animal suffering. *Applied Animal Behaviour Science* 20, 135–142.

RSPCA (1992) *Care for Your Kitten*. Harper Collins Publishers, London.

Rubin, S. (1995) *Practical Guide to Cat Care*. Publications International, Lincolnwood, Illinois.

Sanchez-Andrade, G. and Kendrick, K.M. (2009) The main olfactory system and social learning in mammals. *Behavioural Brain Research* 200, 323–335.

Santiapillai, C. and Supraham, H. (1985) On the status of the leopard cat (*Felis bengalensis*) in Sumatra. *Tigerpaper* 12(3), 8–13.

Scarlett, J.M., Salman, M.D., New, J.G. and Kass, P.H. (2002) The role of veterinary practitioners in reducing dog and cat relinquishments and euthanasia. *Journal of the American Veterinary Medical Association* 220, 306–311.

Schaller, G.B. (1972) *The Serengeti Lion: A Study of Predator–Prey Relations*. University of Chicago Press, Chicago, Illinois.

Schneck, M. and Caravan, J. (1993) *Cat Facts*. Barnes and Noble Books, New York.

Schulkin, J. (1999) *Neuroendocrine: Regulation of Behaviour*. Cambridge University Press, Cambridge, UK.

Schwartz, S. (1998) Animal behavior case of the month. Aggression of a cat toward its owner when cleaning the cat's litter box. *Journal of the American Veterinary Medical Association* 212, 1384–1385.

Schwartz, S. (1999) Use of cyproheptadine to control urine spraying and masturbation in a cat. *Journal of the American Veterinary Medical Association* 214, 369–371.

Schwartz, S. (2002) Separation anxiety syndrome in cats: 136 cases (1991–2000). *Journal of the American Veterinary Medical Association* 220, 1028–1033.

Schwartz, S. (2003) Separation anxiety syndrome in dogs and cats. *Journal of the American Veterinary Medical Association* 222, 1526–1532.

Scottish Government (2010) *Code of Practice for the Welfare of Cats*. St Andrew's House, Edinburgh.

Seager, S.W.J. and Demorest, C.N. (1986) Reproduction of captive wild carnivores. In: Fowler, M.E. (ed.) *Zoo and Wild Animal Medicine*, 2nd edn. W.B. Saunders, Philadelphia, Pennsylvania, pp. 667–706.

Seidensticker, J. and McDougal, C. (1993) Tiger predatory behaviour, ecology and conservation. *Symposium of the Zoological Society of London* 65, 105–125.

Serpell, J.A. (1986) *In the Company of Animals: A Study of Human–Animal Relationships*. Basil Blackwell, Oxford, UK.

Serpell, J.A. (1988) The domestication and history of the cat. In: Turner, D.C. and Bateson, P. (eds) *The Domestic Cat: The Biology of its Behaviour*. Cambridge University Press, Cambridge, UK, pp. 151–158.

Short, C.E. and Van Poznak, H. (1991) *Animal Pain*. Churchill Livingstone, New York.

Siegal, M. (ed.) (1997) *The Cornell Book of Cats: A Comprehensive and Authoritative Medical Reference for Every Cat and Kitten*. Villard Books, New York.

Siegel, A. and Edinger, H.M. (1983) Role of the limbic system in hypothalamically elicited attack behavior. *Neuroscience and Biobehavior* 7, 395–407.

Siegel, A. and Pott, C.B. (1988) Neural substrates of aggression and flight in the cat. *Progress in Neurobiology* 31, 261–283.

Siegel, A. and Schubert, K. (1995) Neurotransmitters regulating feline aggressive behavior. *Review in Neuroscience* 6, 47–61.

Siegel, A., Schubert, K.L and Shaikh, M.B. (1997) Neurotransmitters regulating defensive rage behavior in the cat. *Neuroscience and Biobehavior Review* 21, 733–742.

Siegel, A., Roeling, T.A., Gregg, T.R. and Kruk, M.R. (1999) Neuropharmacology of brain-stimulation-evoked aggression. *Neuroscience and Biobehavior Review* 23, 359–389.

Sigmund, O.H. (ed.) (1979) *The Merck Veterinary Manual*, 5th edn. Merck & Co., Inc., Rahway, New Jersey.

Silveira, L. (1995) Notes on the distribution and natural history of the pampas cat, *Felis colocolo*, in Brazil. *Mammalia* 59, 284–288.

Sisson, S. (1975) *Sisson and Grossman's The Anatomy of the Domestic Animals*, 5th edn. Saunders, Philadelphia, Pennsylvania.

Smallwood, K.S. (1993) Mountain lion vocalizations and hunting behavior. *Southwestern Naturalist* 38, 65–67.

Smithers, R.H.N. (1968) Cat of the Pharaohs. *Animal Kingdom* 61, 16–23.

Smithers, R.H.N. (1978) The serval, *Felis serval* Schreber, 1776. *South African Journal of Wildlife Research* 8, 29–37.

Smithers, R.H.N. (1983) *The Mammals of the Southern African Subregion*. University of Pretoria, Pretoria, South Africa.

Soulsby, Lord E.J.L. and Morton, D. (eds) (2001) Pain: its nature and management in man and animals. *Royal Society of Medicine International Congress Symposium Series* 246, 17–25.

Spain, C.V., Scarlett, J.M. and Houpt, K.A. (2004) Long-term risks and benefits of early-age gonadectomy in cats. *Journal of the American Veterinary Medical Association* 224, 372–379.

Spinka, M., Newberry, R.C. and Bekoff, M. (2001) Mammalian play: training for the unexpected. *Quarterly Review of Biology* 76, 141–168.

Spreadbury, B.R., Musil, K., Musil, J., Kaisner, C. and Kovak, J. (1996) Cougar population characteristics in southeastern British Columbia. *Journal of Wildlife Management* 60, 962–969.

Stahl, P. (1993) Status of the wildcat (*Felis silvestris*) in Western Europe. In: *Proceedings of a Seminar on the Biology and Conservation of the Wildcat* (Felis silvestris). Council of Europe, Strasbourg, France, pp. 16–25.

Stander, P.E. (1992) Foraging dynamics of lions in a semi-arid environment. *Canadian Journal of Zoology* 70, 8–21.

Stander, P.E. and Albon, S.D. (1993) Hunting success of lions in a semi-arid environment. *Symposium of the Zoological Society of London* 65, 127–143.

Stocek, R.F. (1995) The cougar, *Felis concolor*, in the Maritime Provinces. *Canadian Field Naturalist* 109, 19–22.

Strain, G.M. and Myers, L.J. (2004) Hearing and equilibrium. In: Reece, W.O. (ed.) *Dukes' Physiology of Domestic Animals*, 12th edn. Cornell University Press, Ithaca, New York, pp. 852–864.

Stuart, C.T. (1986) Caracal in Africa. *Cat News* 4, 9–10.

Stuart, C.T. and Wilson, V.J. (1988) *The Cats of Southern Africa*. The Chipangali Wildlife Trust, Bulawayo, Zimbabwe.

Sunquist, F. (1997) Where cats and herders mix. *International Wildlife* 27(1), 26–33.

Sunquist, F. (2001) Staying close to home. *International Wildlife* 31(3), 20–29.

Sunquist, M. and Sunquist, F. (2002) *Wild Cats of the World*. University of Chicago Press, Chicago, Illinois.

Sunquist, M.E. (1981) The social organization of tigers (*Panthera tigris*) in Royal Chitawan National Park, Nepal. *Smithsonian Contributions to Zoology* 12, 239–241.

Sunquist, M.E. (1992) The ecology of the ocelot: the importance of incorporating life history traits into conservation plans. In: *Felinos de Venezuela: Biología, ecología y conservación*. Fundación para el Desarrollo de las Ciencias Físicas, Matemáticas y Naturales, Caracas, pp. 117–128.

Sunquist, M.E., Sunquist, F.C. and Daneke, D.E. (1989) Ecological separation in a Venezuelan llanos carnivore community. In: Redford, K.H. and Eisenberg, J.F. (eds) *Advances in Neotropical Mammalogy*. Sandhill Crane Press, Gainesville, Florida, pp. 197–232.

Sunquist, M.E., Karanth, K.U. and Sunquist, F.C. (1999) Ecology, behaviour and resilience of the tiger and its conservation needs. In: Seidensticker, J., Christie, S. and Jackson, P. (eds) *Tiger Conservation in Human-Dominated Landscapes*. Cambridge University Press, Cambridge, UK, pp. 5–18.

Tabor, R. (1983) *The Wild Life of the Domestic Cat*. Arrow Books, London.

Tabor, R. (1997) *Cat Behaviour*. David & Charles, Newton Abbot, UK.

Tewes, M.E. and Everett, D.D. (1986) Status and distribution of the endangered ocelot and jaguarundi in Texas. In: Miller, S.D. and Everett, D.D. (eds) *Cats of the World: Biology, Conservation and Management*. National Wildlife Federation, Washington, DC, pp. 147–158.

Tewes, M.E. and Schmidly, D.J. (1987) The Neotropical felids: jaguar, ocelot, margay, and jaguarundi. In: Novak, M., Baker, J.A., Obbard, M.E. and Malloch, B. (eds) *Wild Furbearer Management and Conservation in North America*. Ontario Trappers Association, Ontario, Canada, pp. 697–711.

Thomas, E.M. (1994) *The Tribe of Tiger*. Simon and Schuster, Inc., New York.

Thornton, K.C. (2011) Saving Africa's big cats. In: *Cats USA 2011 Annual*. BowTie Inc., Irvine, California, pp. 58–67.

Tilson, R.L. and Seal, U.S. (eds) (1987) *Tigers of the World*. Noyes Publications, Park Ridge, New Jersey.

Turner, D.C. (1988) Cat behaviour and the human/cat relationship. *Animalis Familiaris* 3, 16–21.

Turner, D.C. (1991) The ethology of the human–cat relationship. *Swiss Archive for Veterinary Medicine* 133, 63–70.

Turner, D.C. and Bateson, P. (eds) (1988) *The Domestic Cat: The Biology of its Behaviour*. Cambridge University Press, Cambridge, UK.

Turner, D.C. and Meister, O. (1988) Hunting behaviour of the domestic cat. In: Turner, D.C. and Bateson, P. (eds) *The Domestic Cat: The Biology of its Behaviour*. Cambridge University Press, Cambridge, UK, pp. 111–121.

UFAW (1995) *Feral Cats; Suggestions for Control*. Universities Federation for Animal Welfare, Potters Bar, UK.

Vaillant, J. (2010) *The Tiger*. Alfred Knopf, New York.

Van Orsdol, K.G. (1984) Foraging behaviour and hunting success of lions in Queen Elizabeth National Park, Uganda. *African Journal of Ecology* 22, 79–99.

Verstegen, J., Fargetton, X., Donnay, I. and Ectors, F. (1991) An evaluation of medetomidine/ketamine and other drug combinations for anaesthesia in cats. *Veterinary Record* 128(2), 32–35.

Villablanca, J.R. and Olmstead, C.E. (1979) Neurological development of kittens. *Developmental Psychology* 12, 101–127.

Voith, V.L. (1979) Multiple approaches to treating behavior problems. *Modern Veterinary Practice* 60, 651–654.

West, M.J. (1974) Social play in the domestic cat. *American Zoologist* 14, 427–436.

West, M.J. (1979) Play in domestic kittens. In: Cairns, R.B. (ed.) *The Analysis of Social Interactions*. Lawrence Erlbaum, Hillside, New Jersey, pp. 179–193.

Weston, C. (2011) *Lion Country*. Evans Mitchell Books, London.

Wharton, D. and Freeman, H. (1988) The snow leopard in North America: captive breeding under the Species Survival Plan. In: Freeman, H. (ed.) *Proceedings of the Fifth International Snow Leopard Symposium*. International Snow Leopard Trust and Wildlife Institute of India, Bombay, pp. 131–136.

White, J.C. and Mills, D.S. (1997) Efficacy of synthetic feline facial pheromone (F3) analogue (Feliway) for the treatment of chronic non-sexual urine spraying by the domestic cat. In: *Proceedings of the First International Conference on Veterinary Behavioural Medicine*. Universities Federation for Animal Welfare, Potters Bar, UK, p. 242.

Whitney, L.F. (1953) *Cat Care*. Doubleday and Company Inc., New York.

Whitten, W.K. (1985) Vomeronasal organ and the accessory olfactory system. *Applied Animal Behaviour Sciences* 14, 105–109.

Wikipedia (2010) Wikipedia. Available at: http://en.wikipedia.org (accessed 2010).

Worah, S. (1990) Rusty spotted cat *Felis rubiginosa*. *Cat News* 12, 12.

Wright, J.C. and Amoss, R.T. (2004) Prevalence of house soiling and aggression in kittens during the first year after adoption. *Journal of the American Veterinary Medical Association* 224, 1790–1795.

Wyatt, T.D. (2003) *Pheromones and Animal Behaviour*. Cambridge University Press, Cambridge, UK.

Young, L.E. and Jones, R.S. (1990) Clinical observations on medetomidine/ketamine anaesthesia and its antagonism by atipamezole in the cat. *Journal of Small Animal Practice* 31, 221–224.

Index

Note: **bold** page numbers indicate figures; *italic* page numbers indicate tables.